Salad Bar Beef

by

Joel Salatin

Polyface, Inc.
Swoope, Virginia

This publication is designed to provide accurate and authoritative information in regard to the subject matter covered. It is sold with the understanding that the publisher is not engaged in rendering legal, accounting or other professional service. If legal advice or other expert assistance is required, the services of a competent professional person should be sought. -- *From a declaration of principles jointly adopted by a committee of the American Bar Association and a committee of publishers.*

Salad Bar Beef, First Edition
Copyright © 1995 by Joel Salatin

Cover photo by Tom Gettings

Layout by Vicki H. Dunaway

Library of Congress Catalog Card Number: 95-071623

ISBN: 0-9638109-1-X

Contents

Harvesting the Salad Bar (continued)

Maintaining the Salad Bar

Marketing the Salad Bar

Extending the Salad Bar

Appendices

Acknowledgments

Any endeavor like this requires collaborative effort, and my indebtedness to others' skills and perception is huge. A few names bear mention, however, beginning with my own family — wife, Teresa; son, Daniel; and daughter, Rachel — along with my mother, Lucille and posthumously, Dad (William), for their patience during the long days of trial and error, and recently at the computer.

Special thanks go to Roger Wentling, who introduced me to *Stockman Grass Farmer* and the able grass movement catalyst, Allan Nation.

Specifically on this endeavor, I appreciate Robin Leist, whose cottage industry, Keystrokes and Design, helped with computer interfacing and produced the beautiful cover. Finally, many thanks to Vicki Dunaway, whose desktop publishing system, guidance and proofreading margin notations *("Does this really make sense?")* made this book a reality.

Special credit, too, is due all of our "cheerleading" patrons, who bolster us with compliments, teach us with constructive criticism, and deepen our faith in the viability of Salad Bar Beef.

Foreword

Despite today's low cattle prices you can make a good profit with a small beef cattle operation. This book will show you how. Joel's Salad Bar Beef prototype as described is a financially better suited prototype for 95 percent of the cow-calf producers in the United States than the sale of commodity calves or yearlings. However, this is not just a "how-to" book. It is also a book of philosophy, feelings and beliefs. Some may wish that Joel would just stick to the "facts," but for learning to be truly effective it must necessarily be a triad of why, how and who.

"Why" consists of basic principles, observations and deeply held beliefs. "How" is the specific proven responses to specific problems, and "who" is your psychological support group or cheerleading squad. Most of us think we can shorten our learning curve by concentrating on the "how" and ignoring the rest. We do so at our own peril.

Real learning, the type of learning that can make you a living, comes primarily from doing it wrong at first, recognizing it didn't work, modifying our actions and trying again. Without knowing the "why," without having a vision of what a successful prototype should look like and a deeply felt belief that such a prototype is possible, we won't persevere in the face of repeated learning curve "failures."

"Who" is as important as how and why because none of us likes to fail. Without a group of people rooting for us and urging us on, most of us will quit after the first couple of failures. Interestingly, thanks to his direct marketing program, Joel is able to draw upon the warmth and approval of the people he feeds for his psychological support.

"Who" is also the sharing of information with others who have done, or are attempting to do, what we are doing. Without this necessary networking, we are all doomed to repeat each other's mistakes. To that end, this book is invaluable.

Joel's Salad Bar Beef program is a proven, profitable prototype that can make an excellent profit from a small cow herd regardless of the commodity price of calves. While you may want to modify some of the production parts of Joel's prototype to make it mesh better with your particular rainfall and climate, the integrity of the "whole," and in particular the direct marketing portion, must be preserved for it to be as profitable for you as it has been for Joel.

Go for it!

Allan Nation
Editor
The Stockman Grass Farmer magazine

Introduction

In a day when beef is assailed by many environmental organizations and lauded by fast-food chains, a new paradigm to bring reason to this confusion is in order. With farmers leaving the land in droves and plows poised to "reclaim" set-aside acres, it is time to offer an alternative that is both land and farmer friendly.

Beyond that, the salad bar beef production model offers hope to rural communities, to struggling row-crop farmers, and to frustrated beef eaters who do not want to encourage desertification, air and water pollution, environmental degradation and inhumane animal treatment. Because this is a program weighted toward creativity, management, entrepreneurism and observation, it breathes fresh air into farm economics.

The yearning for a comfortable living from a pleasant life in the country is pandemic, as is the desire within folks to reconnect to the land and eat clean, nutritious food. The salad bar beef paradigm speaks to these issues with answers that escape band-aid problem solvers. I do not have all the answers, but this approach offers a positive perspective on a topic that generally receives nothing but negativism, ranting and railing.

It is my hope that this book will challenge the very foundations of America's agriculture and provide some creative alternatives to conventional wisdom. Yesterday's battles over DDT will pale as we enter the era of genetic engineering and more insidious food processing manipulations. The herbivore and grass relationship is an important ecological building block. For the sake of health — ecological, human and societal — the salad bar beef model cries out for acceptance.

The sooner we adopt it, the better.

JOEL F. SALATIN
August, 1995

Considering

Salad Bar Beef

Chapter 1

The Salad Bar Beef Opportunity

"The middleman takes all the profits."

"Everybody wants lean beef, but packers won't pay a premium for it. That doesn't make sense."

"When beef goes down at the sale barn, do you think it goes down in the supermarket? Not on your life! I'd like to get what those retailers get just one time."

And you can. Every livestock producer is used to saying or hearing these refrains. They snap shut like emotional traps on our farms and ranches, causing an adversarial relationship between packer and producer, or producer and retailer, and finally producer and consumer.

It's time to opt out. If conventional marketing channels won't or can't respond to our efforts, give us our price or satisfy consumer desires, then we must redirect our focus to alternatives. We must produce it alternatively and market it alternatively.

Salad bar beef can be produced with literally nothing but some electric fence and walking shoes. It does not require capital-intensive investments like heavy-metal weighted enterprises. Haymaking can be done on a contract basis and there is no tillage, planting or grain harvesting/storage/handling required. This production model is beautiful in its simplicity, both for what it does not require and for what it does require. It requires tremendous flexibility, information and observation, but these things need not be borrowed from a bank.

These are skills, easily transported from where you are to where you might want to be. In fact, it is common to begin this type of enterprise without owning any land. Because it does not require machinery and buildings, it can be started on a small scale and still be profitable. The only reason farms need to be large to be profitable is because they are leveraged toward materials handling and heavy metal. A three-yard tractor bucket moves material like silage or manure cheaper than a one-yard bucket. This is why farms dependent on heavy metal and materials handling need to keep getting bigger and bigger in order to stay in business.

As soon as we eliminate the capital-intensive elements of the farm, its profit potential becomes relatively size neutral. We can rent a five-acre pasture and use some portable electric fence and be in the livestock business. The total investment in fence and water pipe might be a couple hundred dollars. But as soon as we need standard equipment, buildings and row crops or purchased grains to operate that five-acre pasture, its profit potential goes by the wayside. We need 50 or 100 or 500 acres.

Salad bar beef offers existing farmers a way out of their economic morass and offers wanna-be farmers a way into farming that can bring a white collar salary from a pleasant life in the country. Because of the low-money input production model, salad bar beef can be produced cheaper than its grain-fed counterpart. To be sure, it has a much higher informational and observational input, but again, these are generally not big cash costs. Even if the price obtained for the product were the same as sale barn prices, the profit potential is higher because the gross margin per pound produced is better than conventional production models that rely on continuous grazing, grain, fertilizers and heavy metal.

But when we couple lower production costs with a higher sale price through niche marketing, the gross margin becomes large indeed. By wearing the hats of producer, marketer, retailer and advertiser, we pick up the shares normally siphoned off by these parts of the food industry.

On our farm, we average an equivalent liveweight price of a little more than 90 cents a pound, and this includes cut, wrap and freeze costs. Steers average 500-600 pounds carcass weight and heifers are roughly 100 pounds lighter. On average, we gain $200-$300 per head, net, by marketing direct. That is significant.

What is especially interesting about this is that the only difference between these prices and sale barn prices is our marketing effort, which is not subject to weather, disease, pestilence and price. Think about it. What are the four big variables in agriculture? They are weather, disease, pestilence and price. The more our income is derived from sources subject to these variables, the more risky our enterprise. But as we leverage our income toward marketing and processing, we insulate our dollars from the production variables.

In a drought, our customers don't generally move. When grasshoppers come, our customer base stays intact. When prices at the sale barn fall, our prices are at retail levels which seldom fluctuate. As we change the dollars in our income pie from production dollars to marketing and processing dollars, our farm business becomes more stable. If only 20 percent of our income derives from production and 80 percent derives from marketing and processing, only 20 percent of our dollars are subject to these big variables about which farmers sit in coffee shops the world over and complain.

Furthermore, when the variables do come, salvage is more likely because instead of trying to save large volumes of cheaply-priced commodities we need only save small volumes of valuable commodities. Value-adding is a time-proven, effective income-booster for anyone, but especially for flexible, small-scale farmers.

Chasing these dollars does not require equipment, buildings and costly inputs. It's largely done from the telephone, the postal service and personal contacts. These are not big ticket items. Pursuing these additional dollars does not require bank loans and high risk. It can be started with one animal; the payback is size neutral. You don't need a potbelly load to make it pay. Small operators like us have all the more reason to spread the gross margin on each animal; and with a high gross margin we can quickly net as much on a few animals as most folks do on a hundred or more.

American livestock producers, long told how to produce and not how to market, have become so detached from salesmanship that they view consumers as the enemy. People want lean, clean beef. The days are ripe for salad bar beef. Many of our customers are not switching dollars from supermarket meat counters to us. They are adding our beef to their previously non-beef diet. They have refused to eat beef for years and now that they've found what they've been looking for, it's almost like they're trying to make up for lost time. How many American consumers have been alienated from the standard fare because of concerns over the environment, humane animal treatment, toxins and fat?

The beauty of this consumer demand is that it asks for what is in our country's best ecological interest. Livestock producers can make more money than they thought possible, insulate themselves from wholesale price fluctuations, quit planting grains, abandon expensive feedlots and receive praise and compliments from appreciative customers.

While this explanation of the salad bar beef opportunity has focused primarily on the production end, it includes the consumption end. The opportunity exists for people to vote with their food dollar every single day. Every time we eat, we vote. And with salad bar beef, consumers can vote for a completely different paradigm, have a higher quality product on their plate, and know that today they've made a difference. Perhaps you are not interested in the production side. That's fine. But you can present this concept to an acquaintance or relative who could or should join the salad bar beef model. That could be your ministry for this day.

Yes, the salad bar beef opportunity is real, for both producers and consumers. The differences in quality, in economics, in the environment and in quality of life are tremendous, and I trust more and more folks will join together to promote and enjoy this opportunity.

Chapter 2

Why Salad Bar Beef?

Salad bar beef.

Like Romeo and Juliet grappled with "what's in a name" I've grappled with "what's in a name" when deciding what to call our beef. The standard terminology simply wouldn't get the job done:

Grass Finished : Our pastures contain more than grass. We wanted to convey the idea that these animals had a huge variety of forages. Furthermore, I didn't like the idea of "finished" because that indicated we were trying to produce beef comparable to what came out of feedlots, but just do it on grass. No, that wasn't the answer.

Grass Fattened : Again, this had the same problem as the first one, in that grass was a limiting word and did not convey the variety of legumes, herbs and weeds — the polyculture — in the pasture. And the idea of "fat" for sure had negative connotations. Our beef is lean, not only because that's what we want, but because that's what's marketable today.

Range Raised : We liked the idea of openness, of outdoors, conveyed by this title, but it did not address the fresh paddock every day concept. Plenty of people have been disappointed by the tough, gamey taste of range-type beef. In addition, our beef isn't just "raised" outdoors; it is palatable, special.

Non-feedlot beef: This was even worse than any of the above alternatives. We realized quickly that trying to describe our beef by what it was NOT would be an exercise in futility. Besides, a positive approach is always superior to a negative one. If our position can be described in a positive way, we get farther with folks. What we needed was a term that would describe:

• A diet of greens and free of high calorie grains

• A table (buffet) offering multiple menu items over which cattle were free to graze

• A self-helping feeding arrangement (grazing) as opposed to feedlot arrangements where feedstuffs are brought to the animals by machinery and labor

• A fresh plate of unwilted greens every day.

Suddenly the idea dawned on me: we're offering these cows a fresh salad bar every day. They have multiple items of fresh greens over which to graze and they get a fresh buffet every 24 hours. "Salad Bar Beef" was born.

Once we had the name, spin-off implications were multitudinous. Americans today worship at the salad bar. Nearly every doctor in the country is putting folks on salad bar diets. The low calorie, "lite" diet has revolutionized eating habits in the last couple of decades. Practically anyone over 50 who is regularly seeing a physician (why would anyone want to do that?) is on a meat-restricted diet. Meat, and especially beef, epitomizes obesity in the mind of most Americans.

As soon as we tell people we produce "salad bar beef," they see, in the mind's eye, "beef in a salad bar." Of course, that is not what we mean, but that is their perception. From a marketing standpoint, perception is reality. And so the notion of "salad bar beef" makes them think this is an item they may find added to the tomatoes, lettuce and diced turkey in the salad bar. And if it's in a salad bar, it must be okay.

There's something about an item carrying the "salad bar" conceptual tag that okays it in the mind of the average person. It's like a superstar endorsing a pair of tennis shoes. It would be wonderful to arrive at the day when "grain fattened beef" held such negative connotations that no one would dare use the term. But until then, borrowing a term from the restaurant is a great way to heighten awareness and capitalize on the universal agreement that salad bars are good for you.

From a marketing standpoint, the "salad bar beef" idea has some real positive aspects. First, it is unregulated. Labeling laws require a host of documentation and often political gymnastics on the part of producers before they can use certain words—especially "organic." Similar words like "biological" and "natural" carry this baggage as well. But because no one in Washington has thought of this term yet and it does not have a widely accepted usage, it is free from these encumbrances. May it ever be.

Of course, I am not a fan of labels anyway, but these laws too often extend beyond labeling to any public or private discussion, directory notations and explanatory brochures or business cards. Since this term is brand new, it baffles the bureaucrats and piques the curiosity of most folks. And that brings us to the second major benefit from the marketing standpoint.

The term is guaranteed to spawn questions. "Salad bar beef? What in the world is that?" I guarantee that the first time you use the term on someone, it will result in this response. It's happened to me every time I've used it. Everyone wants to believe there's a beef that's special enough, from a dietary standpoint, to pass the salad bar test. People are predisposed to crave such an item, and the thought that such an item just might exist is almost too good to imagine.

Rather than coming across with the proverbial "hard sell," this catchy term makes the sales pitch an educational response to the question. It combines sales with education. Rather than seeing ourselves as selling beef, we see ourselves as teaching about beef.

The first is never as gratifying — for either party — as the second.

The term "salad bar beef" requires a doubletake from any consumer. It sounds too good to be true. It's what people dream about in their deepest desires. It's a revolutionary concept — in a good sense. In a disarmingly positive way, it allows the producer to describe the positive differences between his animal protein and the competition's.

For a lot of reasons, then, the term "salad bar beef" works. It works as a descriptive term and a marketing term. It is positive rather than negative and captures the hearts of many.

Chapter 3

What is Salad Bar Beef?

Salad bar beef is not just grass finished, grass fattened or non-grained beef. It is not necessarily pastured beef, either. When I use the term, I have several fundamental concepts in mind. No term is comprehensive, but this one comes close enough to function; it also conjures up such a strong word picture that the concepts are inherently easy to remember because they flow from the picture.

First of all, salad bar beef is lean. It's not fat — on purpose. The origins of heavily marbled beef in America are twofold. The first reason was for the tallow market. Before electricity, when folks needed burnable animal proteins for lamps and candles, it was important to have fat animals. Anyone who peruses a beef production book of pre-1950 cattle will notice how stocky and fat they are. There was a definite reason for producing this type of beef. Indeed, the market for fat was as big as the market for muscle, so it was clearly wise to produce an animal that would appeal to more than one market.

I do not know if the second aspect — marbling — and its related fast-cooking technique developed independently or as a result of the first reason, but it became, and still is, fairly well accepted that the amount of fat governs the palatability, tenderness, and juiciness of the meat.

Americans are the only people on earth who demand that their beef be summarily incinerated in 5 minutes as standard cooking procedure.

Every other culture in the world and throughout history has slow-cooked meat. From the pot of stew in Biblical Old Testament times to the day-long earth-sheltered cooking of Argentina, in its historical context beef has been cooked long and slow under low heat. We know now that this preserves the nutrition and that fast cooking encapsulates the proteins so that they are much more difficult for the body to metabolize. It's just plain harder to digest fast-cooked, high-heat cooked meat than the lower, longer type. If for no other reason than nutritional, then, we should cook meat long and slow.

Perhaps stir-fry or shishkabobs come as close to acceptable fast-cooking as anything. These two techniques, however, require thin and/or small pieces of meat, reducing the bulk, or mass, of the material. The way the heat goes through it, and its relationship to escaping moisture, differs from large pieces fast-cooked. In the large pieces, the outside is practically burnt before the inside is done. Thin slivers, however, respond differently.

Another reason to cook meat slow is tenderness. Flames harden the proteins and dry out the meat, making it tough and unpalatable. Again, I don't know if what I call incineration cooking developed as a means of getting the fat out of meat or if it simply developed because with all that fat in the meat it could be cooked faster without being unpalatable. But the fact is that heavy marbling does make beef more forgiving in an incinerator-cooking model.

Although lean beef is just as palatable as fat beef, it does often require different cooking techniques. We will discuss some of these techniques later; right now we are concerned only with defining salad bar beef. That definition includes "lean but tender." If it is not tender, then it does not fit the term. The bottom line on any food is how it tastes. The proof is in the eating.

Salad bar beef is tasty. It is not strong or gamey like some wild meat, but it certainly is not bland and tasteless like normal feedlot beef.

Salad bar beef receives no grain — ever. God made multi-stomached animals to eat forages: that's why He gave them multiple stomachs. The reason herbivores fill such a wonderful niche in the web of life is that they convert low-energy forages into extremely nutritious food. Some anthropologists have linked less physical strength and endurance to a vegetarian diet along with a concomitant increase in physical prowess among tribes and cultures that added meat to their diets.

The Pueblo Indian exhibits at the cliff dwellings in Colorado Springs, Colorado point out the extremely small stature and short lifespan of the gardening Pueblos compared to the size and strength of the Apaches in the plains who followed the bison. These meat-eating tribes grew in numbers dramatically after the Spaniards left horses. Horses made hunting more efficient and further increased the nutrition of these tribes.

From an environmental standpoint, it is truly remarkable to realize that millions of acres of grassland, from the Great Plains of North America to the pampas of South America to the steppes of Asia can be utilized in a symbiotic relationship with herbivores to produce some of the most concentrated and nutritious human food available. But feeding herbivores grain violates this natural spot in the food chain. It requires soil-harming practices on a grand scale.

At its most fundamental level, then, salad bar beef is a return to the diet of perennial polycultures that have fed herbivores throughout history. The diet does not require cultivation or mechanical harvest, except for haying, which is a relatively non-invasive technique from an environmental perspective.

The forages are multiple, and ideally perennial. The more the better, for each plant occupies a different stratum above ground as well as below it. The picture of the cow grazing across a salad bar buffet is too graphic to go unmentioned here. The more varieties she can choose, the more apt she will be to satisfy her own needs.

These plants must be fresh, not stale. Obviously, then, the cows must be moved from paddock to paddock, frequently, in order to keep from grazing stale plants. The metaphor of salad bar here may not be entirely accurate, but the pasture goes stale, or wilts, far earlier from the cow's perspective than it does from the human's. Nothing could be more stale than overripe forage, and yet to the untrained human eye, the pasture may look fabulous.

By the same token, an overgrazed pasture is equivalent to empty containers in the salad bar. Just as you or I would mull around the buffet, waiting for someone to fill up the containers before we went to the effort of picking up the empties and summarily dumping the dregs onto our plates, cows will mull around a stale paddock looking for a new fresh morsel rather than eating down onto the soil surface to get another nibble. In fact, a cow will only spend 8 hours a day eating. If she doesn't fill up in that amount of time, she will simply quit, ruminate with what she has, and lose weight.

The paddock then must not only offer variety and freshness, but enough quantity in accessible portions so the stock can find and consume what they need quickly. All of this bespeaks a well-maintained salad bar.

Finally, I think the salad bar beef should be processed and marketed locally. Sometimes this may not be possible, but certainly "locally-grown" and "freshness" in a salad bar are more possible when the ingredients are harvested from local producers. The difference between fresh local produce and that produced from far away and trucked across the country is too obvious to require elaboration. In-season local vegetables always taste superior to out-of-season trucked-in merchandise of the same variety.

And although freezing and packaging for year-round use may not be consistent with our salad bar metaphor, I think the "local and fresh" is a worthy goal, and a noble element to strive for in our salad bar beef modeling. Keeping it local adds a bioregional dimension that may not be necessary for palatability or marketability, but is an important dimension nonetheless.

When the average beef animal sees more of America than the farmer that grew it, the food system is inefficient and unnecessarily energy-consumptive. A different model offers solutions instead of additional problems. Ideally, salad bar beef is perennial polyculture, pasture based, locally-grown, seasonal, and exhibits superior taste and nutrition. Nothing in the production or processing is smelly, inhumane or neighbor-aggravating. The whole model is positive for all who see, smell, hear, touch and taste.

Chapter 4

What is Wrong with the Beef Industry?

While this book is about positive alternatives to the current beef paradigms, it is important as a matter of record to articulate, in fact to justify, why a different approach is necessary. "If it ain't broke, don't fix it" would require that we stay with conventional models unless the current models are "broke." My deep conviction is that the current beef industry is "broke."

At the risk of aligning myself in any way with him, I'll mention that Jeremy Rifkin's *Beyond Beef* offers some compelling evidence for such a notion. At the outset of this discussion, however, let me admit that nothing is a "compelling" argument. Matters like this, regardless of statistical data, research support and "empirical" evidence, eventually find agreement in the human mind ***only*** after finding a home in the human heart.

One of my favorite questions when speaking at alternative agriculture conferences is to ask for a show of hands: "How many of you have ever argued anyone into supporting alternative agriculture?" I've never had a single hand go up. Why?

Because the heart screens what the ears hear. In other words, unless the heart is sympathetic, all the so-called empirical data in the world will not convince the naysayer. I offer this introductory aside, in this chapter, to explain a couple of things. First, I'll be short on hard data and long on philosophy. Anyone who truly wants to know the truth can find plenty of specific information, complete with charts and graphs, supporting the position that the current beef industry needs to be modified.

It is not my purpose to offer such data, but it is certainly available elsewhere. Reinventing the wheel has never been my intent. It is at the conceptual level that we set our course. In other words, where we are headed, conceptually, determines our mode of transportation, the tools we'll need and necessary supplies to get there. Once we've decided to go somewhere, it's too late to ask the question: "Should we really go there?" Too often we are too busy trying to keep the car running, the spare tire patched and the clothes washed to ask the question: "Should we really be taking this trip in the first place?"

For example, I've watched in consternation as the alternative agriculture community spends countless hours and millions of research dollars trying to figure out how to produce "organic" beef in a feedlot. Who will dare ask the question: "Why have a feedlot?" The ability to solve problems is directly related to the questions we ask. In fact, our questions define the bounds of our problem-solving. We will never solve a problem we don't perceive as a problem. And generally our paradigm defines what we perceive as a problem. As a result, our paradigm defines the creativity we'll bring to changing any current model. It's a real dilemma, and only shows that we need lunatics like me to go beyond the current paradigms and ask more probing questions.

In the case of feedlots, for example, as long as the feedlot is our paradigm, grain will be the answer. As long as we're bringing feed to the animals instead of the animals to the feed, grain will win out in a high production system because it's easier to handle and you get more calories per pound handled. The feedlot necessarily requires a mechanistic, heavy metal, high input answer because it's that kind of a model. Allan Nation, editor of *Stockman Grass Farmer*, often jests: "If the only tool you have is a hammer, every problem looks like a nail." How true.

If the only way to finish beef is in a feedlot, the tools are quite limited. It is time to add some tools to the hammer. It is time to suggest that beef can be finished in other ways besides in a feedlot. For that matter, the alternative agriculture community spends a lot of time trying to figure out how to raise organic chickens in a big chicken house. Dare we ask the question: "Why raise them in a chicken house?" As soon as we ask that question, a whole host of viable alternatives to pharmaceuticals, groundwater contamination, *Salmonella* and air pollution come to the fore. Inappropriate models should be junked from the get-go rather than being propped up.

How much time is spent trying to figure out how to prop up flawed paradigms? As soon as we throw out the wrong model and institute a new one, all the problems solve themselves. For example, when we throw out the feedlot model, suddenly we don't have a dust problem, we don't have a manure problem, we don't have a disease (coccidiosis) problem, we don't have a feed transportation and storage problem. These problems ONLY exist because of a flawed production model. As soon as we challenge the basic concept, the conventional paradigm, all the effort currently expended solving these problems becomes completely unnecessary. Imagine what a different world it would be if all creative energies could be focused on REAL problems instead of artificial ones manufactured by flawed paradigms.

To summarize my first reason for being long on concept and short on data, then, let me suggest that the data is an outgrowth of the concept, and not the other way around. We all know that "you can use numbers to prove anything." At the risk of sounding flippant toward research and study, I find myself enjoying data that supports my concepts and generally denying data that is opposed to it.

Which brings me to the second reason I'll be long on concept and short on data: if you don't want to believe what I say, I won't convince you anyway. This is not to belittle anyone or poke fun. I've already admitted I'm as guilty of this as anyone. I'm not accusing anyone of something I haven't already admitted in myself. But if you disagree with me, no amount of data, no amount of "facts", will convince you otherwise and it's a waste of my time and yours to try. A fellow just told me the new slogan of the conventional research community is: "I'll see it when I believe it."

I would hate to waste my time as well as yours, so let's be up front about the issue and go from there. We have many clichés that speak to this point: "You can lead a horse to water, but you can't make him drink." "None is so blind but he who will not see." Allan Nation's favorite is: "Don't try to teach a pig to fly. It wastes your time and irritates the pig." My older brother got me an appropriate penholder for my desk that says: "My mind's made up, don't confuse me with the facts." That's what older brothers are for, right? Anyway, I know my limitations and will not presume that I can any more break through an unbelieving spirit with facts and figures than that a preacher can break through a hard heart with the love of the gospel.

Our hearts define what our minds will believe. Call it spiritual, call it what you want. But it is a fact. And if our hearts are in agreement, this discourse on the problems with the current beef industry will be easy to understand. If, however, you think that how an animal is fed has nothing to do with the quality of its meat, that animals are more mechanistic than biological, that natural relationships are made to be manipulated instead of followed and that the soil is just so much inert material to hold up plants, then we will just disagree and that is that. Nothing I can say will change you, nor you I. Fair enough?

The primary indictment against the beef industry is its premise: herbivores should eat grain. In nature, no herbivore eats grain. This grain eating adds unnecessary fat to the carcass, which translates into grease and fat for people. Grain feeding puts more fat in the muscle (marbling), whereas forage fattening tends to put fat on the outside. The carcass looks more like venison in that it has some backfat, but not marbling.

Roughly 70 percent of all the grain grown in the world goes through multi-stomached animals. Imagine what a different culture, a different economy, it would be if 70 percent of all the acreage currently producing row crops were converted to perennial polycultures? Well-managed pastures of perennial polycultures would build soil fertility, convert carbon dioxide into soil carbon (if there is a "greenhouse effect," this would eliminate it), build soil instead of eroding it, and require far less energy to maintain.

I have to laugh to myself when I hear farmers criticizing the Chicago Board of Trade for its power over agriculture, or the power of equipment manufacturers and multinational corporations that produce chemicals and fertilizers. More often than not, these same farmers are feeding grain to herbivores, or producing grain to feed herbivores. Nothing would pare these powerful organizations down to size faster than to eliminate feedlots and grain feeding of herbivores. That would do more than all the tractorcades and Farm Bills combined.

Grain production is a notoriously energy and soil consumptive activity. Certainly there are some grain farmers who practice good husbandry and my hat is off to them. Growing grain like some of the good practitioners I've seen is truly a wonderful thing, kind of like good poetry. But for the most part, grain production is responsible for the lion's share of soil erosion, chemical pesticide, herbicide and fertilizer usage, and energy usage in agriculture.

Well-managed perennial polycultures, however, especially those with native species adapted to the local climate, do not require tillage, are naturally resistant to weeds, build fertility and require no cultivation or mechanical harvesting to maintain. They are far less prone to drought and therefore require less if any water for irrigation. The figures that show a pound of beef costing hundreds of gallons of water are assuming irrigated grain feeding. As soon as we change the production model, the big environmental arguments against beef fall flat. On many points I agree wholeheartedly with environmentalists: beef, as currently produced, is an ecological disaster.

The answer is not to quit eating beef but rather to patronize beef production models that are environmentally enhancing. We farmers are our own worst enemies in many respects. We complain about the environmentalists, and then practice a type of agriculture that is inherently injurious to the environment. If we would quit doing the wrong things, we would not fuel the fire of the opposition. The only reason such antagonism of beef exists is because we farmers have abused the environment, we have utilized erroneous production models and we are only witnessing the inevitable backlash that occurs by angry people reacting with half knowledge.

One of the most joyous aspects of writing this book is to extend a hand to beef haters and show a way to produce this product in an environmentally enhancing way. It can be done and it is being done, and I trust that we can walk together through this explanation and come out to mutual benefit on the other side of the discovery. I am a commercial farmer ***and*** an environmentalist, and not a pseudo-environmentalist like those who claim that without chemical agriculture millions would starve and we would have to plow down additional acreage to make up the production shortfall. Again, that argument assumes grain feeding of herbivores. Take away the grain, and the computer modeling falls flat.

Herbivores are not as efficient at converting grain to meat as hogs and chickens, and this simply exacerbates the inefficiency of the grain system. Generally, seven pounds of grain put on one pound of beef, but it only takes three pounds of grain to a pound of gain for hogs and two to one for poultry. Beef producers are kidding themselves to think they can compete with pork and poultry by feeding grain. The more the price spread between beef and the other two major competitors, the smaller the market share and the slimmer the gross margins for producers.

The second major problem with the beef industry is what I call loosely the "pharmaceutical farm" concept. This is a catchall for all the needling that goes on in the industry, from vaccinations, which are immuno-depressant, to subtherapeutic antibiotic feeding (creates R-factor pathogens, some transferrable to humans) to routine systemic grubicides and wormers like Ivomectrin.

Since when has man been able to fool nature? "Nature always bats last" is a phrase well worth remembering. These tiny pathogens can go through a myriad of generations in just one year, adapting themselves to new drugs and new chemicals faster than human invention can respond. It is so obvious that these chemical procedures have failed that to try to prove it seems redundant. Pyrethroid fly ear tags brought resistant flies within one year of introduction. Parasites develop immunities routinely and farmers are admonished to use different wormers from time to time to slow this process. Herbicide resistant weeds, pesticide resistant bugs, the list goes on and on. When will we quit kidding ourselves that we can laugh in the face of nature and go our merry way?

Now of course the stakes are getting higher as we indulge in gene splicing, genetic engineering and other ludicrous ventures. Remember when DDT was the next best thing to toilet paper? It's amazing to watch the same phraseology, the same arguments, being advanced for using biotechnology as were advanced for using DDT and its cousins a generation ago. The same boat crosses the same river. Will we never learn?

The tragedy of all this, of course, is that none of these things is the weak link. Their use is just not necessary. We have plenty of food in the world — tons of it rots in ports every year. Anyone who thinks people are starving because there is not enough food simply isn't believing the evidence. People starve because of politics, not production. We absolutely do not need more food.

And even if we did, we wouldn't need biotechnology, or chemicals, for that matter. For example, pastures on average are so mismanaged right now that they could produce three or four times their current levels without any tillage, fertilizer or chemicals. Such an astounding jump would render completely unnecessary all the biotech research that promises piddling 10-30 percent productivity increases. If all the acreage currently in row crops were converted to well-managed salad bar pastures, the total meat protein production would double. That is way more than the numbers bandied about by the gene manipulators.

While I will admit that the dangers to biotech have not been fully proven, by the same token neither has the safety. I've always been taught that it is better to err on the side of safety than err on the side of disaster. There is simply no justification for flirting with disaster, especially when the whole reason — increased production — is not necessary.

In addition to grain being wrong with the beef industry, centralization is a great problem. A centralized food system is inherently problematic. It concentrates animals, and that automatically causes sanitation, husbandry and manure problems.

When animals are concentrated, they are more prone to diseases. Talk to any animal science major or professor at any American university, or any industry animal specialist, or attend any livestock conference and disease is the topic of discussion. It is an all-consuming topic in modern American agriculture because it is impossible to put animals on top of each other in close quarters without sanitation and sickness problems.

A pall of fecal-contaminated dust hangs over western feedyards, where tens of thousands of beeves are confined in corrals for months at a time. This pathogen-laden fecal dust invades the body through the respiratory system. If we tried to house humans in circumstances as unnatural to our normal hygienic requirements we would see equivalent outrage in the world community. We would see disease epidemics.

Husbandry problems compound in concentrated conditions because we simply cannot give the individual attention required to maintain health. As we reach the tolerance threshold for survival in such unsanitary conditions, we can't keep up with it all so we resort to medications and the slaughter plant to minimize losses. In a healthful pasture situation, disease is practically nonexistent and good husbandry can be practiced because almost no individuals need special attention.

And finally manure becomes a major headache. Mountains of manure. This is perhaps one of the greatest tragedies because manure is such an important asset. Instead of being returned to the soil to build fertility for the next generation, it is viewed as a "waste" product and treated like a worthless by-product. There are some large-scale composting enterprises, to be sure, but for the most part this manure vaporizes (many western feedlots can be smelled miles away), leaches (ground and surface water contamination from animal nutrients are a growing hazard) or otherwise is disposed of by burning.

If that manure were spread on green vegetation by the grazing animal, it would be assimilated by the living soil and utilized by plants to capture more solar energy. With huge feedlots, we translocate solar energy from where it is needed and concentrate it where it cannot be utilized. We make up the inequities with petroleum and call ourselves efficient and clever.

Centralization in agriculture causes huge transportation costs. In fact, it takes roughly 15 calories of energy to put 1 calorie of food on the American table. Of those 15 calories, 4 are for transportation. The average T-bone steak sold in American supermarkets has traveled nearly 2,000 miles. How efficient is that?

It makes far more sense to finish the beef where it is conceived and sell it in the same region. Not only does that make for a more stable food system, but it reduces the energy required to move the product from field to consumer. Removing consumers from producers fosters mistrust and misinformation. Keeping everything bioregional fosters a sense of community and mutual responsibility.

In a nutshell, what is wrong with the beef industry is essentially what the anti-beef crowd has been preaching for some time. It relies on "toxic rescue chemistry," as Charles Walters, Jr., editor of *Acres, USA,* says repeatedly in his editorials. But the good news is that it need not be that way. This is a positive book about a positive alternative, and while we may take another little jab or two at the conventional beef industry, the rest of this book will be upbeat, encouraging and proactive.

Chapter 5

Fat Animals, Fat People

"You are what you eat." This trite old adage, commonly accepted, should challenge farmers and consumers alike with the production model used in animal agriculture. I am not a medical doctor, but through various experiences and discussions with those who are, I offer the following argument for radically altering conventional animal production models.

If you go to a doctor with a cholesterol, saturated fat problem, the doctor will finger four basic areas. First, he'll want to know something about your eating patterns and he'll advise you to lay off the high calorie stuff and opt more for fruits and vegetables. He'll encourage you to adopt a high vitamin/mineral diet as opposed to a high calorie diet. Veggies and fruits are higher in vitamins and minerals as a percentage of total weight, than are animal proteins (meat, poultry, dairy), potatoes and pasta.

Next, he'll tell you to get exercise. The link between cardiovascular disease and a sedentary life-style has been so well documented it has practically become a maxim. Fitness centers and health clubs are everywhere. Heart patients are admonished to walk regularly for exercise.

Third, the doctor will encourage you to get fresh air and sunshine. There is something about being cooped up all day in an office or in front of a computer that reduces health. Just sitting in full-spectrum sunshine, breathing fresh non-ducted air stimulates deep breathing and a sense of well-being. "Getting away" has become a fancy way of saying "getting out."

Finally, the doctor, if he's interested in more than just seeing how many times he can swing his door open in a day, will ask you about stress. The link between stress and cholesterol is well documented, to the point that company management personnel attend stress management seminars. Stress can be dietary, although it is not normally. It is generally emotional and spiritual, and we have names for some of these things: like "psychosomatic". The point here is that stress does affect us physically in all kinds of negative ways.

The four general areas, then, are diet, exercise, fresh air and sunshine and stress.

Now, what have we done in American agriculture in the last three or four decades? We've taken our animals off pasture and confined them on high calorie, low vitamin-mineral rations. Blue silos and high moisture corn are the norm for much of beef production. In fact, the diet is so unnatural that if the animals were not slaughtered as soon as they are, they would die. Enlarged livers are the norm. We've eliminated the salad bar in their diet and put them on as high a caloric intake as possible. The poultry industry uses train car loads of grease and fat that are by-products of food processing.

The goal with these rations, of course, is fast, cheap growth. It is time to challenge the assumption that growth is inherently good. Cancer is growth, but it is not healthy growth. Nothing about growth is inherently good; growth in and of itself is not a noble goal. And yet we have an entire industry, from crops to animals to vegetables, bowing at the altar of growth as if growth in and of itself is worthwhile. The bigger the better, the faster the better. We know that running the assembly line too fast results in shoddy workmanship. It is no different in animals. Accelerated growth is unnatural and unhealthy. In nature, all animals consume green material, from birds (poultry) and dogs to herbivores. Some take in more than others as a percentage of their total diet, but a percentage is always there.

Chlorophyll, that wonderful material that makes photosynthesis happen and inhabits all green plants, is a detoxifier. Health food stores for decades have carried tiny chlorophyll capsules, touting them as a natural kind of Roto Rooter, cleaning out and detoxifying the body.

Magnesium, a large component in chlorophyll, is equally important as are the B vitamins. Manganese is there in large quantities. Green material is extremely high in B vitamins, which nutritional researchers have linked to curing everything from insomnia to nervousness to epilepsy. Whole books using both replicated and anecdotal research illustrate the validity of nutritionists' claims.

When mothers sternly tell their children to eat their greens, they aren't forcing the youngsters to take poison. They, as well as our salad society, know the value of green material in the diet.

In addition to taking the salad out of animals' diets, then, we've eliminated the exercise. We've crowded animals into feedlots and factory houses where a sedentary life-style is strictly enforced. Now certainly forcing the animals to walk miles a day to feed and water is not the ideal either. But a happy medium, like what is offered by small pasture paddocks, is a good alternative to the feedlot.

Fresh air and sunshine, the third general area relating to cholesterol and saturated fat, is denied modern beef. The pall of fecal particulate and dust hanging over the average feedlot shows well enough that the animals can't acquire these most basic components for health. The link between excessive nitrogen, especially, and disease, is becoming stronger all the time. The nitrogen-laden fecal particulate inhaled by confined animals is astounding. How this affects the meat is still unclear, but every action has a reaction.

Finally, stress takes its toll. There is an old joke about the feedyard manager who went out to see how the cowboy was getting along checking the pens. It was during the spring thaw period, and all he could see were the cowboy's shoulders and head bobbing through the cattle pen.

"How ya' doin'? Everything alright?" he queried.

"Waaal," drawled the cowboy, "I'm gettin' along alright, but my horse is strugglin' a bit."

Anyone who has visited a feedlot in the winter knows that while this story is a stretch, it has serious undertones. With nowhere to lounge that is clean and dry, the animals stand in mud, often up to their bellies. Tails hang heavy with manure. On concrete, the filth is just as bad. The concrete is hard on their legs and the wet manure builds up and covers their bodies every time they lie down. This all adds up to stress.

I don't want to be accused of being an animal worshipper, and equating animals with humans. But by the same token, animals do have certain behavioral and instinctual needs. For example, work done in California in the 1930s and 1940s showed that chickens have more disease and do not lay as well whenever they are in groups of more than 300 birds. Pre-1950 poultry science books, which by the way are the only ones worth reading, show 1,000-bird laying houses with four partitions, sectioning the birds off in 250-bird quadrants.

I've often wondered how many sicknesses currently plaguing the poultry industry could be eliminated just by partitioning off those huge houses so that only 300 birds were together at a time. The point is that behavioral and instinctual needs must be met in order to eliminate stress on the animal. And just because a cow is lying down chewing its cud does not mean its system is unstressed. Beef producers owe a lot to people like Bud Williams and Burt Smith who have devoted their lives to eliminating the stress in livestock. The thing that so amazes me about their findings is how insidious stress really is. It can be happening and we don't even realize it.

The bottom line here is that the very things that have become almost axiomatic as causing cholesterol and saturated fat problems are the very things we've forced on animals in general and certainly on beef specifically. It is magnified in beef because of naturally being an herbivore. Denying salad to herbivores is far more tumultuous than denying it to poultry or hogs.

Yellow fat on poultry and beef, extremely orange egg yolks and naturally yellow butter reflect high levels of chlorophyll in the diet and low levels of saturated fat. In fact, some nutritional doctors recommend these types of eggs to help detoxify their patients. Cholesterol isn't a problem.

In all the years we've been producing salad bar beef, we've never lost an ounce from hot weight to cold weight even after hanging carcasses for two weeks. In the industry, shrinkage is normally a couple to a few percentage points, and is often noted in contracts. The protein in salad bar beef is pure food, not so much soft energy in white fat.

Pour off the grease from supermarket fare and note how the fluids separate. The saturated fat rises to the top and gets hard. In salad bar beef, the percentage is much lower — almost nonexistent. Pastured poultry does not even have any hard fat on top of the container. It's more like vegetable shortening.

Because of the way we've begun raising animals in our society, is it any wonder that we are experiencing more and more cholesterol, cardiovascular, saturated fat-related problems? Why would anyone think that the procedures that exacerbate these maladies in humans would not have concomitant results in animals? Philosophically and experientially I see a direct relation between the two.

If everything relates to everything, and we are what we eat, then obviously production models that foster cholesterol and saturated fat in people will do the same in animals. Is it a stretch to think that eating those animals, in turn, will simply compound the problem in people? It's a vicious circle. Fortunately, salad bar beef offers a positive answer. Perhaps salad bar beef offers not only a completely different saturated fat profile, but a different vitamin-mineral profile as well. I am unaware of any research to this effect, but it would not surprise me at all to find more magnesium, more B vitamins per pound than in supermarket fare. In fact, I'd bet on it.

Perhaps eating salad bar beef is like vicariously consuming a salad. Now there's a thought to ponder. Good food without guilt.

Developing the Salad Bar

Chapter 6

Family Background

I come from an uninterrupted lineage of farmers. Dad and Mom (William T. and Lucille) came here to the Shenandoah Valley in 1961 after losing a farm in Venezuela during the revolution and political instability of 1959-60.

My great grandfather, "Happy" Smith was an extremely successful farmer in Indiana during the late 1800s and my grandfather, Frederick Salatin, always aspired to farm but was as much a craftsman-tinkerer-inventor as he was a farmer. During the Depression he moved off the home farm ("Happy" Smith was killed when a charge of dynamite accidentally exploded while he was blowing stumps out of a new field). My father, William, was only a child when the family moved to town, but remembered vividly the diversified mid-western farm.

Grandpa was a master gardener, and a charter subscriber to Rodale's *Organic Gardening and Farming* magazine. He raised strawberries, raspberries, grapes and honeybees on his half-acre city lot, supplying local produce markets with his "Fred Salatin" personalized boxes of fruits and vegetables. He also produced eggs from a henhouse located in the corner, and grew sugar beets for the birds' wintertime supplement. His henhouse had nails in the walls about 18 inches off the floor and he would take sugar beets out during the winter and impale them on those nails. The hens loved to peck them off the walls.

He held the patent for the very first walking garden sprinkler. Moving things must be in my genes. He wanted portable water and went to his huge basement shop to invent a contraption to do the job. I remember well his compost pile and neat, lush garden beds, heavily mulched to protect the soil.

Even as a boy, my father wanted to be a farmer. But Grandpa's love for the soil, for portability and for the unconventional, including direct marketing, kept Dad from being a normal farmer or dreaming normal dreams. He wanted to go to a developing country where regulations wouldn't interfere with his plans.

He and Mom purchased a farm in the highlands of Venezuela, cleared some land, and began raising chickens. They built a house and a large free-span garage with attached poultry pens. He took the chickens to the village market and soon the vendors, who in turn peddled the birds on a pole through the streets to their customers, would wait for Dad to arrive at the market. His birds did not have subclinical pneumonia like all the indigenous birds. The vendors would run their finger down the bird's beak to check nasal drip. Their marketing depended on having the best quality birds, and telltale pneumonia was so common Dad had no trouble selling his clean birds.

The goal was to have a broiler and dairy operation.

But it was not to be. We were not missionaries, diplomats or corporate employees. It was the beginning of the period of the ugly American, the capitalist hatred spread by so many Communist-instigated coups. A prime target for the junta leaders, Dad was unwilling to "buy" protection from the local constable so we were forced out of the country, leaving almost everything behind. Dad and Mom started over in their early 40s, along with their children. I was four years old in 1961; my brother Art was seven and my sister, Loretta, was born that very year. We looked at farms from Pennsylvania to North Carolina, wanting to be within a day's drive of Washington D.C. because we were meeting with government officials trying to get a settlement for the Venezuela farm.

Eventually, through the efforts of radio muckraker Drew Pearson, the Venezuelan government gave a token settlement and with that Dad and Mom bought a small herd of not quite 30 Hereford cows. The farm on which we settled, 550 acres located on the western edge of Virginia's Shenandoah Valley near the town of Staunton, had an interesting history of its own. It was the homeplace of a much larger farm that dated back to the late 1700s. A descendant of the original owner had five children and when he died in the late 1800s the siblings could not agree on a settlement and so it was divided in five parts. This farm is two of them, a neighbor has two and another neighbor has the other piece.

Because of the split-up and these two pieces going outside the family, what is now our farm was owned by absentee landlords for several decades. It was sold on the courthouse steps in a sheriff's sale around 1915. It was tenant farmed until roughly 1948, when a moneyed family from New Mexico purchased it and salvaged the old farmhouse, which was in such disrepair that cows and sheep were using it for a barn. They stayed for four years and sold it to a man who had struck it rich in Middle Eastern oil ventures, which were just beginning at that time.

Israel was just becoming a nation then and clans were clashing over landholdings much like claim-jumping in America's western expansion. This man owed money to a fellow and that fellow came looking for him. He found him here and acquired the farm. By now it was 1957, four years after the New Mexicans had moved on. Dad and Mom purchased the farm from the second oil man from the Middle East in 1961. Being a very conservative farming community, neighbors began talking about this place that had had three four-year owners, one from New Mexico and two from the Middle East. And here comes a family from Venezuela!

It was the era of McCarthyism, and all these "foreigners" (no matter that all of us were Americans) led the community to believe this farm was the hub of some international spy ring. We awoke one morning to see that the Klan had burned a bale in the lane overnight. It is hilarious now, but I think all this shows our family clearly has a legacy of lunacy and false perception by others.

Anyway, we bought a farm that had been abused for centuries, especially by so many tenant farmers. We measured 14-foot deep gullies. Large areas were solid shale or limestone outcroppings. Those areas grew corn and small grain during the first century and a half of settlement. The Shenandoah Valley is 30 miles wide and 100 miles long, receiving about 30 inches of rainfall a year. Hemmed in by mountains on either side, it was the tall grass prairie of the eastern United States. Early settlers wrote of riding their horses through the grass that was tall enough to tie in knots above the horses' saddles.

Labeled "The Breadbasket of the Confederacy" by historians, the valley produced grain until the west opened up with railroads and plows. It is now primarily a livestock area and practically all annual cropping goes into silage.

Within the first year, Dad realized the farm could either pay a mortgage or salary, but not both. He brought in both private and public agricultural advisors who counseled him to build silos, plant corn, and graze the forest. He eschewed all that advice on both economic and environmental grounds, took outside employment, and began the years of conservation to heal up the erosion. Mom too worked out and they finally paid off the mortgage in 1969.

Dad was very much a visionary and inventor, developing a portable electric fence system in the early 1960s. He built a portable veal calf barn with slatted floor and four quadrants. The milk cows would shade up near it and graze next to it in the pasture. At milking time, the calves would jump outside, suckle the nurse cow tied to a corner of the veal trailer, and then go back in for some fresh hay and grain. This procedure kept the calves close to the cows, spread the droppings in the pasture, and gave the cows a comfortable place to lounge.

He read some literature on controlled grazing and knew how valuable rest and control were. He remarked to neighbors how some year he would have a cow per acre. Of course, everyone thought he was nuts, but today, as we approach that, it does not seem so far-fetched. Too bad Dad did not live long enough to see it now. Not much has changed, though—everyone still thinks we're nuts. We reforested about 60 acres, built some ponds, and began the slow process of land restoration.

Soil was so thin in many areas that Dad poured concrete in old tires and sunk two pieces of half inch pipe in the concrete: one straight up and down and one at about a 10 degree angle for slopes. We could insert half-inch electric fence stakes in these and get them to stand up on the soilless slopes. We put up loose hay with an old hay loader and Dad designed and built a 20-foot trailer that, using cables and winch, would dump clear up beyond vertical. The idea was to stack loose hay like bread loaves on end, in the barn and eliminate the grapple fork. The machine worked, but the hay didn't. As it approached vertical, the top hay, which was most loosely stacked, would slough off and tumble down. Instead of a nice 20-foot bread loaf, we had a 10-foot pyramid, and the idea had to be abandoned. At that point we joined the twentieth century and bought a square baler.

But the years of loose hay had necessitated efficient shed feeding of hay since it was too time-consuming to transport loose hay out to a field, and Dad designed a portable V-slotted feeder gate that we could move through the hay as the cows ate. A series of holes in the barn wall offered one-foot advancement increments and the system was extremely efficient for feeding — and sanitary.

As early as 1964, Dad was brainstorming about portable shade to go along with portable fencing and rotational grazing. He wrote to editors at *Progressive Farmer* and just about anywhere he thought he could get some counsel, to design a shademobile. It finally became a reality in the early 1970s. Meanwhile, pastures were slowly beginning to improve. I can remember as a teen mowing hay and losing my place because the grass was so thin. The 100 acres scarcely supported 15 cows in those days.

In the mid-1970s we installed several miles of permanent electric fencing and tightened up the paddock shifts from every couple of weeks to less than a week. The difference in production was dramatic. At the same time, we acquired a chipper and began composting the manure in the hay shed from winter feeding.

The combination of tighter controlled grazing, compost, and shademobile to spread pasture droppings brought fertility up dramatically. During the early 1970s, as a teenager, I had a flock of laying hens in portable pens and sold eggs at a local farmers' market, called the "Curb Market." Begun during the depression as a way for farm families to earn cash, by the time I began selling there in 1970 as a 13-year-old it had dwindled to only two vendors: two elderly women. I joined 4-H in order to sell there, and that exempted us from all inspection requirements. We were under the auspices of the Extension Service, which had arranged the exemption with the state inspection department.

One lady sold baked goods (that's where I discovered my affinity for pound cake and potato salad with sweet pickles in it). The other lady came from a more diversified homestead and sold pork, baked goods and vegetables. At the time, we were milking a couple of Guernsey cows by hand. We sold eggs, beef, pork, rabbit, chicken, vegetables, butter, cottage cheese and buttermilk. The market opened at 6 a.m. on Saturdays and closed about 11 a.m.

The Curb Market experience taught me marketing skills and business acumen. Dad taught me early the need to make a profit, so that I could replace my inventory of hens as they wore out. I developed a great appreciation for consumers, for selling a product that exceeds the customers' expectations, and for competitive pricing. That was in the days before health awareness hit the mainstream, when people were still eating butter and beef. Farm chemical usage hadn't even become a vogue topic yet. I also learned how much more money can be made by value adding to the product and taking it directly to the customer.

Dad viewed this as an excellent opportunity to build a direct market business free of encumbering inspection requirements, providing higher quality food than conventionally-produced food, but at a competitive price. When I went to college, we shut the booth down, I liquidated my 300 laying hens, and unknowingly, hammered the last nail in the curb market coffin.

Four years later, the market was gone. Apparently, the two elderly matrons had been holding on, inspired by the enthusiasm of a youngster. While classmates were doing their thing, I was up at 4:30 on Saturday mornings to load the car and have my booth open by 6 a.m. People were waiting in line when I got there, too.

When he began working off the farm, Dad liquidated the Hereford cow herd and gradually built back by buying baby calves and feeding them on the bottle with extra milk from our couple of family milk cows. We were up to a dozen cows by the time the mortgage got low enough to finish , so Dad liquidated the herd again. He hated indebtedness. We built back again the same way.

After graduating from college, I returned home and worked as a news reporter for the local daily newspaper — but my heart was on the farm. Roughly two years later, Sept. 24, 1982, I made the break and returned full-time to the farm. Teresa and I had married and we fixed an apartment in the attic of the farmhouse so we could live cheaply. We called it our penthouse. Dad and Mom paid for the farm and gave us a running start. We took that foundation and continued to refine and develop until it now has become lucrative.

Teresa and I drove an old clunker car, never ate out, lived on a couple hundred dollars a month, and committed ourselves to a dream. People who want to farm must be committed enough to sacrifice for it. A lot of people want to be where we are now, but it did not come overnight or easily. We shortened the grazing to one-day stays, built more ponds, installed better water systems, increased the composting, added poultry, began relationship marketing, added pork, sold more firewood, and watched in amazement as the thin soils deepened, the rock areas grassed over, the weeds left and legumes moved in. It was like a miracle happening before our very eyes.

Truly, God's hand has been on our endeavors, and although He has never been early, neither has He ever been late. We've never gone wanting and it has been an adventure we would wish on anyone willing to risk all to commit themselves to a vision. My question to anyone desiring to start farming is this: "What are you doing right now to make it happen?"

Too often would-be farmers complain about no opportunities when they spend their free time tinkering with the hobby car or watching videos. They could devote their time and attention to a nearby farmer, working for free if necessary, to show their character and commitment, to merit being entrusted with a farm at some future date.

Opportunities still exist for those willing to go against conventional thought and commit themselves to a dream. I trust that this book will inspire you in new directions, with renewed commitment. And if you think it may be too late for you, perhaps it's not too late for your children. The only thing worse than starting late is starting never. We certainly don't have all the answers, and we have a long, long list of refinements we still want to make, but the path of discovery is fabulous. How privileged I have been to walk down that path.

Chapter 7

Getting Started

Getting started in salad bar beef is as varied as the number of people who start. I can't begin to offer the definitive formula for everyone. In fact, anyone who thinks this book is a template, and that it can be used as a pattern to cut out duplicate operations is asking too much of any book. The best we can do is stimulate creative thinking, to encourage folks to dream, and to have a few principles to help shed some light along the way.

With a healthy appreciation on how varied the starting places are, then, let's brainstorm some of the possibilities and minimal requirements for getting started. First, you need to read. The resource list at the back of this book is a good place to start. But there is no teacher like experience, so let's not belabor the academic side.

Second, you need a land base from which to operate. If you already have access to land through ownership, relationship or rental, then look at it from a new perspective. Imagine fields of salad rather than row crops. Imagine getting rid of the tillage and grain planting, harvesting, storage equipment. For some, this will require some serious concentration because grain production has become synonymous with "farming." I've done conference presentations and had an old farmer come up to me, arms folded across his chest, and ask: "Now let me get this right. You don't do any plowing?"

"No, sir, I don't."

"You don't grow any grain at all—just grass. Did I hear that right?"

"Yes, sir, that's what I said."

"Well, then, sonny, you don't really do ***any*** farming, do you?" The statement-question explodes with the realization of discovery, kind of like a cat pouncing on a mouse.

"Well, no sir, I guess I don't do any farming." What can I say?

Anyway, my point here is that instead of thinking first about equipment and buildings, we need to think first about just land and grass plants. People always try to complicate things; that's the bent of humanity. We take simplicity and complicate it. Something as simple as monogamous marriage is turned into divorce and emotional carcasses. Something as simple as soil fertility is turned into chemical fertilizers from sophisticated factories. Something as simple as grazing herbivores is turned into tillage, grain farming, feedlots and trucks.

It's as important, as a person getting started, to think in terms of what you ***don't*** need as it is to think about what you ***do*** need. Perhaps it would be well to make a list titled "What I ***do not*** need." The easiest thing in the world is to go out and buy the latest genetics, tools, equipment, soil amendments and species. The hardest thing is to fully capitalize on the resources we already have. So the first thing in getting started is to ***not*** complicate the simple model of land, grass, and animals.

Land can be acquired in many ways and that topic could be a whole book in and of itself. You can manage a farm, be a caretaker for an elderly couple, work-share with an elderly farmer, rent, buy or squat. Not really, but you get the drift. There are a million ways to get access to land.

Once you've picked a spot, you need to get animals. But I can hear folks saying: "Wait a minute. Don't we need to plant grass on that land first? Don't we need to pull soil samples and fertilize first?"

I did a consultation job once on a farm where a fellow wanted to run cattle. It was an abandoned piece of property, fairly weedy, with some good spots but mostly overgrown. The government agronomy expert came in and said the first thing he needed to do was apply two tons of lime per acre. Oh, really? He owned not one cow and had no fencing except a perimeter.

Although his soil may not have been the best, the farm could still support more than 100 cow-days per acre. If he was going to spend any money, he needed to acquire some cattle first, not buy lime. When soil fertility became the weak link (when the cattle needed to eat more feed than the farm could produce under its current agronomic constraints) then and ***only*** then should he apply lime. The first thing is to get animals on there and harvest what is there, to start nutrient cycling, herd effect, decomposition and income.

Type of cattle again is variable. The main goal here is to gain experience producing healthy, happy animals. You can start with stockers, cull cows, cow-calf pairs, bred cows, baby calves on the bottle, dairy heifers, beef heifers or even board bulls for people. To produce salad bar beef, your long-range goal should be to move toward more and more control of the animal. The more of the animal's life and genetics you control, the more consistently you can produce the kind of product you want.

The problem with direct marketing animals you buy in as stockers is that not all of them will perform equally under a forage-based program. But certainly there are sources with high predictability on forage programs and as long as you keep a consistent source, the outcome should be equally similar. Because we've started over a couple of times with bottle-fed baby calves, our genetics early on were extremely diverse. But with each passing year we have fewer of what I call "embarrassment" animals.

Once you have the animals, you can begin to see what kinds of grasses they like and which ones they dislike. You can begin to see which grasses perform well in your area and which do not. The whole idea of farm-individualized species, both in plants and animals, is foreign to commercial agriculture. I am going to great lengths to make the case for starting with land and animals — ***PERIOD***. I am much more concerned about people getting started with too much baggage than not enough. You can always buy in more as needed, but you can't "unbuy" unnecessary purchases.

I know I've risked disappointing everyone who thought this chapter would give a complete list of start-up "things." But simplicity is elegance, and I want folks to start in elegance. The main hindrance to an effective start is too many encumbrances. Nature is forgiving, but bankers and laboratories and equipment dealers are not. We can make mistakes with animals and grass, and they will recuperate and flex. So start with nature; start with simplicity, and you'll save yourself the hassle of trying to run the race encumbered with leg irons.

Chapter 8

What Kind of Pasture?

I am often asked about what forage species will make the best pasture. After all, the salad bar ingredients are extremely important. Put in the wrong ingredients, and business will go south in a minute.

Again, I have a strange lunatic-type answer: "Whatever is in the side ditches of the road out in front of your house."

My reason is simple: those species are best suited, most adapted, to your area. They thrive without fertilization, without irrigation, and essentially without planting. They are there; they've probably been there for some time, and they will probably be there next year and the next.

At the outset, let me address the "new, improved variety" argument. Virtually every exotic variety I've ever seen requires extra attention of some sort. It needs to be coddled. I have no problem with providing extra care if the benefits outweigh the risks. But often the extra attention can become the tail that wags the proverbial dog.

For example, bermudagrass was developed as a nitrogen-sucker. It was developed because confinement operations had a serious manure handling problem. Native pastures would only uptake 150-200 pounds of N and that was not enough. Confinement poultry houses in Alabama and Georgia, confinement dairies in northern

Florida and southern Georgia, plus confinement hog facilities throughout the south necessitated a plant that would uptake huge amounts of nitrogen.

Furthermore, the chemical fertilizer industry, which worships at the altar of synthetic nitrogen, wanted to encourage a species that required liberal amounts. After all, that's just good business. If I can produce a product with planned obsolescence, and then sell you the upgrade, or the widget to keep it going, that's a capitalist's dream. I have you hooked to my sales and you have to keep coming back and back and back.

As a result of these forces, bermudagrass was developed and it will metabolize up to 800 pounds of nitrogen per acre per year. Certainly the potential to produce beef and milk on such a forage is incredible. It will definitely outproduce anything that is native in the area. But if you do not have a confinement house with a manure disposal problem, it may not be right for you. In order to keep it productive, you would have to apply hundreds of pounds of nitrogen per year, and at a cost of eight to ten cents per pound, that could become expensive if you had to buy it.

I'm not saying bermudagrass should be outlawed. What I am saying is that just because a forage can be advertised as exceptionally productive does not mean it is exceptionally profitable. It all comes back to that bigger is better mentality. We all know that production does not equal profits, notwithstanding three decades of USDA pulpit-pounding to the contrary. The laws of physics that say for every action there is an equal but opposite reaction are not canceled in biology.

Another good illustration is endophyte-free fescue. In my opinion, the whole debate surrounding fungus-infested fescue came about as a result of mismanagement. Had graziers all along been managing their pastures for diversity and succulent, vegetative growth the whole issue would be moot. Research now shows that fescue toxicity becomes a problem when it is a pure stand (no clovers, weeds or forbs) and especially on overmature, rank growth.

This often occurs in the middle of the summer because under

continuous grazing, the legumes and forbs are grazed out so that the pasture becomes a stand of pure fescue. At least, as far as the cows are concerned, fescue is the only thing available to eat. The cows have been avoiding the fescue all late spring because it was less palatable than the legumes, and this means that now that they have to graze the fescue, it is over-mature and rank. So is the problem toxicity, or unmanaged grazing? Another one of these cases where millions of research dollars attacked a symptom instead of a cause.

Anyway, fungus-free fescue was introduced but the stands would die out in about three years and require replanting. Anyone familiar with standard fescue varieties knows that it is one persistent grass. It will survive and thrive when other species die. Now we know that the fungus and fescue have a symbiotic relationship, and the fungus is what gives fescue its tenacity. Well, wonder of wonders.

It is incredible how we are bent on complicating the simple. It is so much simpler just to manage the grazing correctly and enjoy the species God put there. But we can never be content to throttle our cleverness, and try to circumvent what is elegantly simple.

Some of the New Zealand varieties of grasses, too, exhibit this same requirement. When I hear the New Zealanders talking about having to fly on minerals with an airplane, I scratch my head and wonder what they are thinking. Can we actually have sustainable farms, self-regenerating farms, when we must fly on nutrients to maintain productivity? Who is kidding whom?

Please do not take my comments here as a blanket indictment on new seeds. In some situations, certainly even these high-nutrient varieties are good. Certainly planting native species and legumes to increase diversity is commendable, and we've frost seeded some bird's-foot trefoil, some alfalfa and intend to do more—even with some exotics.

But we are in the refining stages, and I certainly would not recommend such practices for beginners. Our staple is still fescue, orchardgrass, timothy, bluegrass, switchgrass, Dallis grass, fall panicum, lespedeza, red clover, white clover, plantain and other forbs.

If you have johnsongrass consider yourself fortunate. I wish we had some. We've had a few sprigs in the past and the cattle go nuts over it. If you are converting from tilled ground to pasture, I recommend seeding down a variety of grasses and clovers—one of the only times you'll hear me recommend asking your extension agent. But he can recommend good varieties for your area.

Generally speaking, though, if you have worn-out pastures or over-rested, neglected pastures, I recommend getting started with good grazing management and nutrient cycling, letting succession kick in and bring you the most adaptable species. Nature knows best what to bring on in your pastures. Every single day hundreds of seeds drop on every acre of ground. They are carried on bird feathers, the wind and coats of mammals.

Nature has a magical way of revegetating the landscape with what is suited to it. If you will but provide the right environment for the germination of the well-suited varieties, you will soon have your salad bar.

André Voisin, the godfather of graziering, in his classic, *Grass Productivity*, documents the amazing succession of forage species after proper grazing management begins. A similar study was done here at the Virginia Middleburg Experiment Station back in the 1950s. Two contiguous pastures were used: one was continuously grazed and the other was divided in six paddocks and the cattle rotated on a weekly schedule (a very lazy rotation by anyone's standards). After five years, the percentage of legumes in the continuous-grazed pasture halved and the weeds doubled.

In the rotationally-grazed pasture the legumes doubled and the weeds halved. The pastures were treated exactly the same and had the same type and number of cattle. Just by resting the fields and applying more grazing pressure at any one time, natural succession selected for higher quality, more palatable species.

Legume seeds will lie dormant but viable for decades — some scientists say centuries — until the environment is right for them to germinate. Then they will sprout. We have clovers where only broomsedge used to be — without planting the clover. The field that was poorest when we came to the farm in 1961 was nothing but dewberries, strawberries, hawkweed and broomsedge. You could not find a clover plant and the grass was poor and thin.

We picked dewberries by the bucketful. Today, that same field is lush with red and white clover, thick grasses and a healthy smattering of forbs. It used to grow thistles so thick it looked like a snowstorm when we baled hay. Now you can scarcely find a dozen thistles on the entire farm. We never planted a seed, sprayed, or applied an ounce of commercial fertilizer.

We controlled the grazing, applied compost, and let succession kick in. It is one of the most powerful forces in nature. No amount of human cleverness can out-compete the power of succession, the tendency of a landscape in which all the animals and plants and nutrients are managed symbiotically, to progress to the next higher echelon of plant community.

What kind of pasture? Whatever you have. Until you fully utilize the pasture you have, do not plant a seed in the ground. It is ludicrous to go plant exotic seeds when you are continuously grazing, or moving the animals only once a week. Until you are moving every 24 hours, and allotting exactly the right amount of forage on each paddock shift, I encourage you to leave your money in the bank and watch what succession will do. It is far more important to properly manage whatever pasture you have available right now than it is to spend yourself poor trying to convert it into what you think might be better.

Chapter 9

Biodiversity

Variety is the spice of life. That old song applies to ecosystems as much as it applies to people's lives. That's one of the most interesting aspects of biological farming. Natural principles apply consistently throughout nature and life.

For example, if all the members of a club, church group or organization believed exactly the same way about everything, it would be easy for that group to go off on a tangent. But the fact that each person comes from a different background, and a little different perspective, lends stability and balance to the group.

The principle works the same way on the landscape. Sameness discourages stability. In nature, stabilizing influences come from plant and animal diversity, where bird eats bug, buzzard eats carcass and weed covers bare soil. Just as there is more going on inside a healthy person's body than may meet the eye, so a healthy, functioning farm simply is a manifestation of many actions and reactions going on behind the scenes.

Rather than seeing monoculture and sameness on a landscape, we should see polyculture and diversity. The diversity of flora and fauna increases at the intersection of any two environments. Generally speaking, the three different environments are open land, forestland and water. Biologists describe the edge effects from these intersections, noting that these areas provide the greatest diversity of plant and animal life.

One of the reasons is that many species require two environments to survive. More species require a combination of two environments than only one.

Birds, which are God's natural pesticides, require shelter and cover for nesting and resting. But for feeding, many require more open spaces. Certainly some birds thrive in mature hardwoods for their entire existence. And some, like meadowlarks and goldfinches, thrive quite nicely in the pasture without any wooded area around.

But these exceptions notwithstanding, there are more species that require open/water, open/forest, or forest/water combinations to thrive. Bluebirds, for example, prey on insects on open land but nest in wood on the edge. In fact, most important insect eating bird species will not eat more than 200 yards from woody cover.

They will fly over an area larger than that on their way to feed, but will not settle down and eat grasshoppers or turn over cow patties to eat parasite larvae more than 200 yards from cover. For some, that cover can be a 20 foot hedgerow and for others, it needs to be a healthy forestal ecosystem.

Photo 9-1. *Biodiversity, edge effect and keylines. Note the permanent electric fence posts along the edge of the field, running on the keyline. Field/forest clearly delineated and brush edge encouraged.*

We've planted wooded areas over the years so that no open ground (pasture) is more than 200 yards from forest. The result has been a plethora of bluebirds, turkeys, mocking birds and crows. I know many people wouldn't like crows, but they do a wonderful job of flipping over cow patties in mid-winter and scratching them all over the place. In the process of spreading out the dung pats, they consume parasite grubs hiding underneath and debug the soil surface. This is all part and parcel of maintaining the hygiene and replenishing the salad bar.

If fertility is relatively high in an abandoned open area, locust trees will come in as a pioneer species rather than conifers. Locusts are the number one habitat for ladybugs, one of nature's great pest controllers. Locusts usually grow most abundantly in a pioneer species situation. In the Midwest, Osage orange serves a similar function.

What that means is that in order to sustain such a species, planned harvesting along field edges is necessary. This opens the ground for locust sprouting. In 30 years the locusts can be harvested for fence posts and the cycle repeated.

It is a travesty of modern agriculture that farms of 500 acres must buy fenceposts and firewood. In the eastern United States, I believe any farm should be roughly 25 percent forested just to have adequate edge effect. In the plains areas, certainly this may be too high. But even in the grassy plains, forested areas exist and the general concept of ecosystem diversity holds true. Exactly what value this contributes to salad bar beef is hard to quantify, but the product is only as good as the whole system.

Pest control is just one aspect of diversity. Another is watershed and water cycling protection. On farms where steep hillsides are reforested, for example, springs begin to run where they had previously dried up. Trees store water, slowly releasing it to the soil and air. In fact, one average-sized deciduous tree cools the atmosphere as much as the air conditioner for an average American house. Bare soil, in contrast, radiates heat and warms the atmosphere. Clean tillage row cropping stimulates desertification faster than any procedure in modern agriculture's arsenal of nature-fighting tools.

It is important to remember that every single desert in the world is man-made. And in fact, deserts will green up again when proper grazing occurs. The Greek historian Homer, author of the *Iliad* and the *Odyssey*, walked across the entire north of Africa, what is now known as the Sahara, and said he never left the shade of a tree. Deforestation to build Carthage, overgrazing and tillage changed the relationship between the earth and sky. The earth, instead of being cool and acting magnetically to make water want to condense on it, so to speak, instead heated up and radiated heat back into the atmosphere. The clouds were still there, even as they are over deserts today, but the rain would not fall.

Research in India's Rajputana Desert shows that by eliminating improper grazing, the earth's surface will revegetate and cool, encouraging rains to fall. Salad bar beef, from a worldview perspective, offers a tremendous benefit and solutions on a grand scale.

Water, which can include man-made ponds as well as those riparian areas that nature builds, is another asset that must be en-

Photo 9-2. *Biodiversity, ponds and protected riparian areas. Fenced out ponds provide aquatic environments in field and forestal areas. Note cattails, brush and trees right down to water's edge.*

couraged. Members of the martin family of birds, for example, drink from still ponds when feeding. Watching the acrobatics of barn swallows while mowing hay never ceases to amaze me. They seem to relish the opportunity to show off, coming inches away from my head and the tractor exhaust pipe. After several successful bug raids, they fly over to a nearby pond and skim the surface, only to return and perform some more.

Frogs and toads certainly eat many garden and crop bugs. Any fish in the pond eat grasshoppers by the score who are hapless enough to jump into the water. Water heats and cools slowly, providing a heat sink in cold snaps and tempering the air during hot periods.

Just as cattle can destroy a riparian area, they can destroy a forest too. They can easily destroy its leaf litter, its regenerating growth, the trees' root collars and, subsequently, those assets like bird habitat and lumber quality. For the farm to function as a healthy whole, care must be given to manage each aspect in a regenerative way, so that there are no losers. All subsets must be winners.

Fluctuating the grazing/haying paddocks from year to year stimulates diversity in the pasture. Because the grazing rotation is done based on grass growth rather than by the calendar, spring grazing begins at the same grass green-up point each year. By starting one year on paddock A, for example, and next year on paddock D and the following year on paddock G, the plants will not be grazed at the same physiological maturity point each year. In addition to the fast daily rotation, the long multi-year rotation allows different paddocks to reach different levels of physiological expression at different times throughout the growing season. This way cool season forages are not favored over warm season, or grasses over legumes, or legumes over herbs. Of course, one of the best techniques for insuring full expression is to let a paddock grow until it's made for hay. One year a hayfield can be made early in the season and the next it can be grazed three or four times and mowed late in the season. Such fluctuations stimulate a variety of seeds being produced as well.

Encouraging the variety — the number of offerings in the salad bar — is one of the best things a grazier can do to insure healthy cows. A polyculture will always justify the time and effort expended managing and planning for it.

When functioning harmoniously, all this multilayered variety acts synergistically. Every livestock producer should desire this synergism. It is the backbone of an efficient, functional piece of land, forest and water. All three ecosystems are distinct yet interdependent. Truly, variety is the spice of life.

Chapter 10

Water

I like ponds.

Certainly different parts of the country have different resources and liabilities, but the pond concept should stimulate some creative thinking.

First, let's identify what characteristics on a farm or ranch lend themselves to building a pond.

Nobody knows your land like you do. I've never been enamored with taking government cost-share money for water development, primarily because by the time the bureaucrats get done you pay just as much for the extra gingerbread as you would pay for the entire project if function were the primary consideration. If you have a seep, or a wet spot, you probably have a pond site. I know there are many other considerations and if you want to know all about them, buy a pond book.

If you don't have a wet spot, but rather have a drainage area funneled into a swale, you can catch runoff during wet seasons and meter it out during dry times. If you have trouble with a pond holding water, get a few pigs and throw them shelled corn on the ground, leaving them to make bricks out of whatever soil is there. Old-timers talk about this method working time and again for leaky ponds.

The point is that pond sites exist all over the place, especially in non-brittle environments. Louis Bromfield was a pond fan, viewing catchment ponds as the answer to flooding.

We've built ponds that only catch surface runoff, and even though it may take all winter for it to fill, that water will last throughout the summer. Such a pond does not need a drain, stand pipe or exit pipe. Just a well-sodded spillway will work fine. Always build the pond deep rather than wide. That reduces surface evaporation and conserves the water through the summer. We always tell the trackloader operator to dig down as far as he can.

Another pond we put in was in a seep. We had a backhoe operator dig it for about $35. It's about 8 feet deep and 12 feet in diameter at the top — cone-shaped. Much of the water seeps in from underneath, but it holds about 3,000 gallons. It only goes dry during extreme droughts, and we just sidestep that paddock during those periods. That's all part of the paddock rotation plan.

The point here is that there is no one way to build a pond, no one place and no one tool. They don't have to be big to be effective. We've built nearly a dozen ponds, and each is a little different. Some have pipes and some don't. All are different shapes. Some have dams, and some are just holes in the ground.

In our area, a good trackhoe operator can build a 250,000 gallon pond in a little over a day at a cost of $600-$800. That's a lot of water.

A pond that size, figuring that a cow will drink 15 gallons on a hot day, will hold 16,000 cow-days worth of water. Any rainfall that does come, of course, replenishes part of the water throughout the year.

In our experience, high pressure systems are sophisticated and costly. You can either buy high pressure or high storage capacity. These are kind of opposite ideas, but essentially if you put your money in high volume storage, you can have low pressure, or extremely simple pumping systems. If you put your money in high pressure pumping and piping, you need no storage.

A ram pump only functions in certain situations: where you have excess flowing water. Wells here cost anywhere from $3,000 to whatever. Pipe and fittings, pressure tanks and pumps, don't come

Photo 10-1. *Simple water system: 80 psi ¾-inch polyethylene black pipe and a 'T' every 200 feet. Extremely cheap system, using regular 99-cent garden hose valves.*

cheap. The point here is that any system will cost money. I think the system that requires the least maintenance and provides the greatest benefits per dollar is the way to go. Certainly ponds cost money to install, but these other alternatives come with a price tag too.

We move the water from the pond to a portable trough a couple of different ways. One is with a 12 volt bilge pump. These high volume, low pressure bilge pumps have gone through some revolutionary engineering in the last 10 years. We have tried different makes, and found the ones manufactured by Rule to be the best.

These pumps are designed to be used in boats, to move water over the hull and back out into the lake. They are high volume, low pressure pumps. We like the Rule because its impeller and outlet port are one molded unit.

The pumps that have a snap-on outlet port allow too much slippage. The molded design insures closer tolerances around the impeller.

Right now the best one we have is a Rule 3700 gallon-per-hour model. It will pump well at 12 feet of lift, but much over that

Photo 10-2. *Gravity fed from a pond more than a mile away, there's plenty of pressure on this system. Gravity overcomes friction in the small pipe, delivering plenty of water on demand.*

Photo 10-3. *A 30-gallon water trough and six-dollar 300 gph float valve with garden hose makes an extremely portable water system. Note the float valve protected under the electric fence wire.*

Photo 10-4. *A 3500 gph 12-volt bilge pump plumbed onto 1¼ pipe.*

shuts it down. We use a 1¼ inch black plastic pipe about 75 feet long to get the water out into a portable tank. The power source is a 12-volt deep cycle marine battery. We keep two so that we can charge one up while using the other one. One charge will run the pump for more than 5 hours. At 3,000 gph, that's 15,000 gallons of water. We put the pump leads on the battery terminals with alligator clamps. These pumps will not lift or push water very much: remember, they are designed to move a lot of water over the side of a boat so your feet stay dry.

One major problem with these pumps is that they use strange-sized outlet ports. I think it's a conspiracy to make sure you buy the special flimsy, flexible hose that marinas sell.

We used a regular 1¼ inch plastic pipe coupling on the outlet port of our Rule 2000, pressing it on over the port. On the big Rule 3700, the port is larger and we finally ended up slitting the 1¼ inch plastic pipe and putting a piece of inner tube behind the flange on the end of the outlet port. Using a regular pipe clamp, we were able to make it snug enough to hold fine.

Since these pumps are low pressure, you need not have tight couplings or fittings. They won't blow off unless they are real loose. The beauty is that once you have a pond, you have a totally portable water system for $100 in a pump, $30 in pipe, $100 in two batteries and $100 in a tank.

We rigged one up with a switch and relay to run a float switch and turn the pump on and off as water was needed in the tank. We watered 100 head of cattle from a 50-gallon trough with this gizmo and it worked fine. Of course, most of the time we use a 350-gallon tank, or two tanks totalling 400 gallons, and just fill them once a day. When I go move the cows, I just switch on the pump and let it run while I move the cows and put up the next cross fence. Before I'm done, the tanks are full and I know the cattle water is fine for the next 24 hours. This is especially effective in the winter, when I fill the tank in the afternoon while the sun is out, and let the system drain back before freezing overnight. Of course, when filling once a day we don't use the float switch or relay assembly.

Photo 10-5. *Getting ready to drop the bilge pump into the pond. Note the vegetation growing right down to the water surface because an electric fence protects pond edges from cattle impaction.*

Photo 10-6. *Dropping the submersible bilge pump into the pond.*

Photo 10-7. *Hooking the alligator clamps onto the deep cycle battery to power the bilge pump. Pump, battery and pipe cost under $200, alllowing riparian protection and full benefit from ponds and aquatic environments. The battery charge will last for about 7 hours of pumping.*

Photo 10-8. *Stock water tanks outside the pond area ready to receive pond water. These high volume, low pressure pumps do not push water far or high, but work well at under 100 feet horizontal and 15 feet vertical.*

Photo 10-9. *Note the clean pond water coming into the tank. While it fills, we can lay out another cross-fence, enjoy talking to the cows, or read a book.*

Since these pumps are made for boats where clean water is the norm, they will not last indefinitely receiving the abuse we mete out dragging it around the fields and in the back of the pickup truck and plopping it in the pond weeds. But even if you have to replace the pump every other year, that's only $50 per year for pumping. I usually place the pump in the water upside down so the screen is not sitting flat on the pond bottom. That way it is pumping clean water.

Certainly the bilge pump is not the answer for every situation, but if you are trying to lift water from a pond or creek out into the paddock, it is a wonderful tool. It enables you to protect the riparian area, which is essential for clean water, healthy livestock and efficient nutrient cycling.

Another option utilizing ponds is to gravity feed water. We built a pond at the bottom of a ravine with a seep in it. The pond is scarcely as large as a swimming pool, but the ravine is a high one, giving us some head pressure. We ran a ¾ inch black polyethylene 80 psi pipe from that pond. The pipe lies on the ground underneath the permanent electric fence and every 200 feet is a 'T' with a little one dollar plastic garden hose ball valve at the end. All the fittings for ¾ inch pipe automatically mate to regular garden hose fittings. We dug a trench under the sod at gates so we could drive cattle and machinery over it.

Because gravity is in its favor, the small pipe gains pressure rather than losing it the farther it gets away from the pond. Even though we run it up high knolls, the pressure stays fine. Using a 50-foot piece of 5/8 inch garden hose and a simple 300 gallon-per-hour float valve on the side of a 40-gallon trough, we have accessible, 50-pound pressure water anywhere on the farm. No pumps, no pressure tanks. Extremely simple.

For another field, we had a spring-fed swampy area below the field and we wanted the water up along the edge of the field. We persuaded our neighbor (where the spring is located — he has it piped into a concrete stock tank) to let us help him fence out his swampy overflow area to keep the water clean. It crossed under the boundary fence and formed into a little stream. We took a shovel and made an 18-inch high dam in the stream, running our ¾ inch pipe through the bottom of the dam and packing the clay all around the pipe. We attached an elbow and a 1-inch nipple about 6 inches long with a cap on it. We drilled inlet holes in the nipple and cap, then covered it with fly screen to keep debris out of the pipe.

We ran this pipe down the bottom of the little stream channel until we had enough head pressure to come up over the side of the channel and begin heading toward the field. Sure enough, we had plenty of head pressure up along the field, and repeated the 'T' every 200 feet. It works great.

In a location where it is impossible to excavate a pond high enough on the farm to gravity feed water everywhere, pumping may be necessary. But even that option has a low cost model. Instead of putting in large diameter pipe (1 inch or higher) all over the farm, put the large pipe in a straight line from the pump to a reservoir at the top of a hill. The reason large diameter pipes are used in a pumping and pressure tank situation is because of friction. As water travels through the pipe, friction reduces the pressure. You can lay a pipe virtually horizontal and after a certain length, there just will not be any water at the end of the pipe.

The idea is to use the big (expensive) pipe for the shortest distance possible, and then use the small, cheap pipe for all the meaderings over the farm. By using the big pipe straight up a hill to a 400-2,000 gallon tank, for example, cheap pipe can come from the reservoir and gravity feed all the paddocks. In many cases, this can be done with a few hundred feet of big pipe. Then the small pipe is the delivery system, but gravity is in its favor so friction becomes a non-issue. A float valve on the reservoir tank is all that is necessary.

If the long arms of the water delivery system come down from the source instead of up from the source, there is absolutely no reason to use big pipe. The price difference between even ¾ inch and 1 inch pipe is amazing. Furthermore, with the reservoir/gravity system, the flow from the source need not be high because it has all 24 hours to recuperate from the momentary use surges throughout the daylight hours. A flow of only 1 gallon per minute will yield

1,440 gallons per day. But if you need to deliver 10 gallons per minute uphill to an on-demand stock watering tank, the delivery system needs to be much bigger, and necessarily more expensive. But delivering 10 gallons per minute downhill from a reservoir can be done with ¾ inch pipe.

The piping required to deliver water to 200 cows a mile from the source is completely different if the source is pushing it downhill instead of pushing it uphill. The savings on large diameter pipe can go into other projects, or more pipe to better meet the needs of the paddocks. This system drastically reduces the typical cost of a water network.

While these options will not exist in every area, they or something similar will work in many if not most areas. The goal here is to think simple, think creatively, and think in terms of least landscaping, the fewest bells and whistles, as ecologically as possible.

Chapter 11

Pond Advantages

Ponds offer some advantages over other water systems. If the cost of pond installation is close to the cost of installing a well or high-pressure/piping system, then a pond certainly should be considered.

All these advantages presuppose that livestock is fenced out — completely. No little notches or V's for drinking, unless it is a big pond or lake that can handle the mud and excrement. As a rule, ponds should be fenced. Period.

Having established that, let's itemize some advantages. First, the aquatic environment is an asset, fostering biodiversity. Remember, the greater the variety of plant and animal life, the greater the stability. And whenever the three environments of open land, water, and forest intersect, we have the greatest variety of flora and fauna. Injecting the aquatic environment in a backyard or a farm encourages the balance biodiversity affords.

For example, ponds encourage frogs and toads, which are great insect predators. Birds that eat bugs are drawn to ponds. Cattails and other hydrological plants filter out toxins from surface runoff.

Because water changes temperature more slowly than air, a pond offers a microclimate of buffered temperature. On the leeward edge, for example, fruits prone to freezing can be grown. The air nearby will be cooler in summer and warmer in winter.

Wildlife is always drawn to water. In fact, during a drought, a pond can mean the difference between death and life for certain species. Of what benefit is wildlife on a farm besides aesthetics? It's hard to quantify, but the wildlife component does offer stability to the plant and animal community. It's all part of that fresh, tasty salad bar.

This is an area where environmentalists and farmers (not that these two need to be mutually exclusive) often lock horns. Imbalances show up in a surprising number of ways. Coyotes killing baby calves is symptomatic of two things: not enough wildlife like voles, moles, field mice and fawns; unseasonable calving. When calves drop in January and February, when coyotes are especially hungry, baby calves make more tempting meals than if the calves are dropping while deer are fawning and wild rabbits are running around and groundhogs are finished hibernating.

When we as farmers are experiencing a problem, the tendency is to blame anyone or anything but ourselves and our management, our model, for the problem. True, it may not be of our doing, but more often than not we can respond more appropriately than pointing fingers.

Balance is a wonderful thing on a farm: open land balanced with forestland; mammals balanced with reptiles; water balanced with field; carnivores balanced with herbivores. Documenting and quantifying the importance and economics of this balance is practically impossible with single-variable, linear scientific methodology, but the benefits are no less important, and no less real.

If the pond can support fish, it offers recreational opportunities. Certainly a large pond, say an acre of surface, could allow some caged aquaculture opportunities. But much smaller ponds can be stocked with bluegills, catfish and bass. An acre-foot of water will grow far more animal protein than an acre of land. One of the reasons is that the water supports the weight of the animal, reducing energy required to stand upright against gravity. It takes much less energy to move 1,000 pounds of animal through the water than over the land—if you don't believe me, try dragging a dead cow out of the barnyard.

In addition to the inherent advantages offered by an aquatic site, ponds help control flooding and erosion. By providing a still place for surface runoff to stop in its downward plunge to creeks and rivers, the stiller pond water allows vital sediment and nutrients to settle out and remain on the farm.

If the dam is built higher than the spill pipe, a degree of flood control occurs by holding back a certain quantity of water and releasing it slowly through a pipe. Every minute and every gallon of water that can be held back during a flood reduces the catastrophe just a little bit.

Fertilizer production is another advantage of ponds. Over time, algae, salamanders, cattails and other plants and animals live and die, falling to the bottom of the pond. This organic material creates a muck. This black, sticky material is perhaps one of the richest fertilizers available. It combines sediment, which is the finest soil particles, and the highly nitrogenous plant and animal life.

We have spread muck on pastures and seen mind-boggling explosions in fertility. I believe if a farmer would build enough ponds to drain and clean one out each year on perhaps a 15- or 20-year rotation, ponds could be the foundation of a fertility program. Certainly that opportunity is not afforded by alternative water systems. Nothing produces sweeter grass than mucked soil.

Another distinct advantage is in water protection. Here in the East, especially, water contamination is becoming more frequent. I know farmers who cannot water their livestock in creeks that traverse their farms because of mercury or nitrate contamination. But if you have a watershed that you own completely or share with a conscientious neighbor, then you can catch that water and know that it is free of contaminants.

Groundwater is certainly not exempt from contaminants. More and more wells are showing nitrate toxicity, especially. But a healthy community of plants around the edge of a pond will filter out nitrogen so that your pond's water can be more sanitary for the livestock than well water or creek water. The point is that if you control your watershed, even if it's only two acres, at least you know the source of that water in the pond and can gain control over the quality of water your livestock drinks.

Finally, ponds always increase property value. I hesitate to mention this one because it presupposes that you may sell the farm, and I don't want to plant that idea in anybody's head (at least not anybody neat enough to build ponds). But because we are not married to the land, that option must be examined. And the fact is that ponds always make land more valuable.

According to real estate figures, the value always increases far more than the cost of pond construction. Part of the reason is that a well-constructed pond gets more beautiful and functional over time. As the water-loving plants begin growing around the edges, the cottonwoods or willows spring up, and the habitat converts to the oasis-type, the value escalates.

Whether we pump out of the pond, gravity feed out of the pond, or siphon out of the pond, the pond has advantages that pressure tanks, wells and underground piping do not. In some cases, underground piping, wells and pressure tanks are the best option. But primarily because much more is written about the higher-tech methods than ponds I have chosen to focus on ponds and simpler systems. We've opted for ponds and never regretted a single one.

Chapter 12

Shade

The shade issue is a divisive one among graziers. As I see it, the debate centers around what is necessary. Certainly some areas have more need than others.

In Colorado, where much of the land is at 7,000 feet up and summer days are cool at night but not more than a dry 90° during the day, shade is not nearly as important as it is in humid areas where the temperature seems hotter.

I will not debate the point about whether or not cattle ***need*** shade. Clearly you or I would not necessarily die if we had no shade on a hot, humid day. But we would certainly not be comfortable. This is the sticky point. Is it enough to give the animals only what they need, or should we give them what will make them more comfortable?

Beyond that, can we change the genetics of the animals to change their comfort level? Most European breeds are most comfortable between 40 and 60 degrees Fahrenheit. Brahman cattle are most comfortable between 60 and 85 degrees. Clearly genetics can play a role in comfort.

If we do not have shade, we would want to breed for summer comfort. You can tell when an animal is not comfortable. When they're panting, salivating, milling around, rest assured that they are highly stressed. Remember that the key to good salad bar beef is lack of stress. What I am suggesting is that while the animals may not need shade to live, or even to be productive, they do need shade to be comfortable. Survivability has never been my goal; comfort is.

I do believe we can virtually eliminate the need for shade through proper breeding. By constantly culling for heat tolerance, we can definitely change the comfort zone of the herd. This is one reason we use a little Brahman in our animals. We can fudge in winter because we use the hay shed for winter feeding.

If I were in Maine, I certainly would not use any Brahman. We need to match the animals to the environment.

But for the sake of comfort, let's address the shade issue. The main problems I see with shade are nutrient translocation and pathogen incubation. Cattle concentrate their pasture droppings under shade. We want those pasture droppings out in the pasture. Concentration of droppings anywhere, whether it is a nighttime campsite, a daytime shade area or a watering hole, represents translocation of nutrients. The cows are taking nutrients acquired from and rightly belonging to the pasture and dumping them in areas where they are either not needed or underutilized.

This is why if we have shade trees in a pasture we want to run our paddock cross fences within inches of the tree, to spread the cattle out from the tree and not let them follow the shade around the tree as the sun moves across the sky. Ideally our bisecting of the tree will be into the sun, not at cross angles. If we give the cows either an eastern portion or western portion, they will spread out more than if we give them a southern and northern portion. Also, if the tree branches are pruned up to 20 feet off the ground, the shade will be away from the tree instead of directly under it. This keeps the cattle from bunching up right at the trunk. They follow the shade during the day and cover much more area during their lounge time.

But beyond translocation, the shade area becomes an incubator for pathogenic organisms. So great a manure and urine load is dumped in this spot that it does not have time to sanitize before the next grazing time. Furthermore, because the area is constantly shaded, pathogens tend to stay alive longer than they do out in the exposed pasture. Sunshine is the best sanitizer in the world.

If we are going to raise clean salad bar beef, we must protect it from pathogen infestation; we must try to keep the animals lounging on fresh linens, so to speak. An alternative model will not work if we just eliminate the things we do not like. If we are not going to use toxic systemic wormers, we must institute positive modeling to compensate for the vacuum. It is not enough to just quit the negative; we must implement the positive. Certainly controlled grazing is the single largest step in the right direction.

But if we need shade to maintain comfort, then it ought to be portable. Dad designed and built a 1,000 square-foot portable loafing shelter. It is 50 feet long and 20 feet wide, big enough for 50 cow-calf pairs. We can move it from paddock to paddock with the cattle, providing them a clean place to lounge every day and spreading out their pasture droppings where we want them instead of where the cows find it convenient.

Photo 12-1. *Cattle lounging under the shademobile, a portable 50 foot x 20 foot tricycle shed that offers a clean shade spot each day. It spreads pasture droppings as well.*

Photo 12-2. *Getting ready to move the shademobile. Tractor is hooked up and we're raising stabilizers on either side of the front.*

Called a "shademobile," it is an integral part of the fertility program and livestock hygiene. Starting with a 10 foot X 50 foot mobile home frame, we beefed it up with 'I' beams and junior beams, added 6 inch steel pipes for legs and built it on three points. The tricycle allows it to be turned 360 degrees. A tie-rod on the two rear wheels allows turning of the rear end also, like big city fire trucks. Quite maneuverable, it is stabilized by two car jacks on the outside of the front that swing up to move, and down after a move.

Two chocks chained to the back wheels ride a couple of inches away and provide a safety when we're pulling the shademobile up a hill. They also save on time on and off the tractor when we park it. We can pull it to the next spot, let it roll back against the chocks until the hitch pin loosens, and then get off the tractor and unhitch it. We do not use it as much as we used to now that we have more heat-tolerant cows, but it is an extremely useful tool and certainly keeps the cattle comfortable on hot summer days.

When we build a second one, the only major change will be to use nursery shade cloth instead of steel roofing. Shade cloth can

Photo 12-3. Now it's on to the next paddock for tomorrow's lounge spot. The shademobile can accommodate 50 cows and their calves.

be purchased according to percentage of light emittance: 20 percent, 30 percent, even 80 percent. Because it is permeable, wind can go through it, which reduces likelihood of wind damage.

This material is far cheaper and certainly easier to install. The undergirdings could be lighter, further reducing weight. The wheel supports, however, need to be strong enough to withstand cow rubbing.

The "savannah" idea has both merit and drawback. A savannah is essentially a grass ecosystem but with trees widely spaced. This is a darling of the double-cropping folks who point out that we can grow a crop of walnuts, for example, as a second crop over the pasture. This results in a two-tiered production model that generates more income per acre. Favorites for the trees, of course, are ones that can take livestock impaction. Most high quality hardwoods do not thrive under these conditions.

Pioneer species like locust and pine perform well under this system. It is widely espoused by followers of Russell Smith's *Tree Crops* and Bill Mollison's *Permaculture.* While I very much practice and appreciate much of what these men say, the logistical nightmare of mowing hay in a savannah, of constantly maneuvering around obstacles in the field, is a real efficiency problem. Trying to get hay to cure under trees, making good hay where some of it is exposed to sunlight and some of it is shaded (the one will get too dry and the other still not be dry enough) has led me to the treeless field concept with the shade problem addressed with a portable tree.

The uneven growth of the grass even affects grazing because in a shaded area the grass will grow one speed while in the unshaded area it will grow another. This again discourages efficient grazing of the paddock. What you have are multiple microclimates throughout a paddock, and while I certainly appreciate this diversity, in this application it militates against efficient field management. This is why we've identified "field" and "forest," and though they are integrated extensively for biodiversity and edge effect, the electric fence divides the two so that we get the benefits of a healthy forest, of biodiversity, without the liabilities of field clutter.

Where we want to grow trees, a double- or triple-tiered approach has more merit. Grape vines underneath walnuts, for example, do not unnecessarily clutter the landscape with obstacles. The trade-off between mobility and diversity is one that each of us must address on the landscape. This is the tug-of-war between the square foot gardening concept and the clean row tillage. Multiple plants in a tiny plot produces more material per surface area, but it must be hand-handled. If we were still mowing hay with a scythe, cutting little spots each day, the savannah may be an excellent technique. But the application of diversity should not eliminate efficiency to the point of making us lose our profit.

Everything must be kept in balance. When our taxes and insurance drop to reasonable levels, then perhaps we can park the tractor more and do everything by hand again. But right now the cash demands of our culture require a certain volume of production, and that volume requires mobility efficiency, grazing efficiency and harvesting efficiency. All these affect quality of the hay, quality of the pasture, and stress on the livestock.

A shademobile is not essential to producing salad bar beef, but each part of the puzzle adds to the whole. People always ask me: “Do I have to do composting? Do I have to have a hay shed? Do I have to have a shademobile? Do I have to have an eggmobile? “

I always answer: “Look, I don’t know how essential each component in our system is to the quality of the beef. I know you can have marketable salad bar beef without all these components, because we started without all of them. But I can tell you that the quality of the beef has gotten better with each incremental refinement, whether it’s the shademobile, hay shed, culling for parasites, or the eggmobile.”

We’ve thought too long about what is the minimum; what can we do to “get by,” like a C student just trying to make that passing grade. I encourage all of us to see the A+, the really excellent, superb performance. Don’t be satisfied with applause: go for the standing ovation.

Chapter 13

Choosing a Breed

There is more difference within breeds than from breed to breed.

Within the Angus breed, for example, some animals are heat tolerant and some look for a shady tree or water hole at the first sign of morning sun. Some Angus are parasite resistant and some are constantly wormy. You can walk into a herd of Angus cows and some will be free of face flies and others' faces will be covered. Some will have a gentle temperament and others will scarcely stay in the corral.

Not only are animals within breeds different, but climates certainly are different. Some breeds do better in a hot environment and some do better in a cold one. Some do better in a dry or brittle environment and others do better in a non-brittle, or temperate region. It is not my intent to provide a laundry list of breeds and all their strengths and weaknesses. Suffice it to say that no one breed has all the assets.

For example, the American Brahman has the following assets: heat tolerance, longevity, browsing ability, parasite resistance, maternal aggressiveness and walking ability. But these long-eared beef animals have the following liabilities: late fertility, slower growing, cold intolerance and mediocre cutability (dress-out percentage). The same set of plus and minus characteristics could be given for virtually any breed. No one breed "has it all." Carefully studying breed association materials can help, but nothing beats talking to producers and looking at comparisons.

Whenever someone says they've got the answer to the beef breeds, rest assured they don't know what they're talking about or they're being hyper-arrogant. Some of these extremely heavily-muscled continental breeds lack maternal aggressiveness, foraging ability and longevity. We must be careful not to unwarily swallow someone's cure-all. Generally an asset in one area will have an off-setting liability in another performance area.

I think the ultimate goal is to find an animal "that works for me." We are back to individualized genetics. Obviously this can't be done overnight, but by watching our animals we can gradually cull to the type that work under our program. It is important to realize that most of the genetic work being done currently is leveraged toward the feedlot approach. Beef is being bred to eat from a creep feeder in the pasture and then go on hot feed as early as possible. Fast-growing beef that weighs 1,200 pounds at 14 months is what the industry is after.

But this type of animal will not perform on grass. Perhaps some of the best thinking on this topic has come from Tom Lassater, who developed the Beefmaster breed. He did not care what color they were; all he wanted was performance. But here again, what performed for him will not necessarily perform for me. It is not my intent to disparage any breed or approve any breed. The fact is that the genetic selection for one breed will not insure success in another part of the country.

We use a three-way cross: Brahman, Angus and Shorthorn. For us, it works well. But I can guarantee that if I were in Vermont, I wouldn't use any Brahman. By the same token, if I were putting these animals in a feedlot, they wouldn't begin to compete with the exotics. But I am competing only with me, not anybody else. And as long as my patrons are happier and happier with their beef every year, that's all that matters.

So how do we know what "works for us?" If we're going to

have an individualized breed, a farm-specific breed, if you will, what is our selection criteria? Hold onto your hat, because this is lunacy at its most basic.

First, we select for fertility. Game biologists know that fertility in a group of wild animals is the first indicator of health. Reproduction occurs when everything else is satisfied. It is the most fundamental function of any living thing; and the first function to slide when things go awry. Incidentally, sperm counts in modern men are astronomically lower than they were a couple of decades ago. Infertility among couples is an increasing problem. Are we noticing?

We want that cow to carry a calf every single year, calve on schedule, and cycle back early. Hard breeders, slow breeders have no place around here.

Secondly, we select for mothering instincts. This includes having an unassisted calf every single year. Again, Lassater was adamant about this, and even culled a cow whose calf was struck by lightning because it gave her an unfair advantage over the others. I don't go that far, but his point is well taken. Every time we give that proverbial "second chance," every time we get overly attached to a cow when she's messed up, we regret it later. I used to ask the vet when we had a breech birth or prolapsed uterus: "Well, what are the chances she'll do this next year?"

His response was always: "No greater than any other cow in the herd." I used to believe that, but not anymore. Now her chances of a duplicate mess-up are much less, because she's hamburger long before next year. Dr. Ike Eller, longtime beef specialist from Virginia Tech, used to tell farmers to cull for "the three O's: open, old, and ornery." That's pretty good advice.

Thirdly, we select for parasite resistance. Back in the early 1970s when we stopped using grubicides and conventional wormers we would run the yearlings through the headgate in the spring to squeeze out heel fly warbles. The mature cows were resistant and we never had any trouble with them getting heel fly warbles. But the calves were another matter. These parasites hatch out in the early spring as tiny flies that lay eggs on the hair follicles on the back of the leg near the hoof. This is the critter that makes cattle run with their tails up in the air. If you watch closely, you can actually see the heel flies down around the ankles.

The egg hatches into a tiny worm that crawls down the hair follicle, burrows through the hide and starts its 12-month migration up through the body and finally out onto the back right under the hide in late winter. By this time it's a one-inch worm. It secretes juice to dissolve a little breathing hole in the hide and continues to grow until early spring, when it crawls out of the breathing hole, rolls out on the ground, and in a couple of days pupates into a fly and the whole cycle starts over.

We would put the yearlings through the headgate before turning them out to pasture and physically squeeze these warbles out — like giant zits — and feed them to the chickens. In a group of calves, we would have some that had only one or two and others that had 15. Same farm, same bull, same field, same group, but incredibly different susceptibility to heel flies. Within a couple of years we had calves that had none. We began culling for this trait and it taught us that the resistant ones were always the ones that were clean and shiny and the parasite-prone calves were also the ones that looked scruffy, had trouble holding flesh, had manure on their tails and were generally more flighty. What a coincidence!

Let me ask a stupid question. How do we know which animals are parasite-resistant if we routinely administer Ivomec and cover up the genetic propensity? The only way we can select for genetic propensities is to allow unimpeded genetic expression in that area. The same holds true for vaccines, medications, fly repellants, ionophores, etc. The more we manipulate with unnatural methods — including grain — the less we know about the hardiness of the animal. I would rather have a group of calves naturally free of parasites than one that required pharmaceuticals and headgating to keep productive — regardless of weight differences.

When the pharmaceutical companies tout: "Get 30 extra pounds of gain with Product X. It'll pay for itself," they are pushing farmers into two traps. First of all, the farmer is denied the ability to find out the natural tendencies of his stock. These pharmaceuticals, as I call them, are very addictive. Once you're on, it takes a bigger and bigger dose to get the same kick and withdrawal is so terrible you stay on just to keep from facing withdrawal. I've found a direct link between these artificial crutches and the need to use more of them.

The more a herd depends on them, the more it needs them. Those of us in the natural livestock system are familiar with hot-house animals that fall apart when they are denied the crutches. These crutches include grain, by the way.

The second tragedy of these ads is that they mislead folks into thinking they will work like this all the time or that these crutches are better than the real thing. You know how the research is done. Some company provides seed money to a land grant university to do some research on the experimental farm. The herd on the experimental farm, of course, is on all the latest crutches available to the industry. That's just the nature of the system. I don't know of any experimental land-grant farm that operates a salad bar beef program.

What that means is that these animals come into the study crutch-dependent already. It's kind of like the research that showed organic farming couldn't produce anything. Typical research went something like this. They would go out behind the soil or crop sciences lab and pick a couple of plots they'd been using for two decades to run chemical fertilizer, pesticide and herbicide tests. They wouldn't do anything to these plots at all: these were the organic plots. In the adjacent plots they would pour on the regular dose of chemicals, or, as Charles Walters, Jr. calls it, "toxic rescue chemistry."

Then they'd plant the corn and at the end of the season take their measurements. Well of course the chemical plot did great and the organic plots did miserably. Conclusion: "If farming went organic either half the world would starve or we'd have to plow up twice as much acreage for the same production. Clearly, organic farming doesn't work."

Such research, of course, makes a mockery of true science. Anyone involved in organic farming knows that it takes a lifetime to build a healthy soil. To take soil that has been abused for decades, do nothing positive to it one year and call that "organic" is a lie. And yet this technique was and is employed all over the world to disprove alternatives.

Ready for another example? Several years ago a chemical plant growth regulator was introduced that kept grass vegetative. The idea was that the longer we delay physiological maturity of forages, the more production we will get and the more palatable they will stay. We get increased production and increased average daily gain on production stock. So research trials compared a sprayed pasture and its animal performance to an adjacent non-sprayed pasture. Both were continuously grazed. Of course, the sprayed one performed much better.

After hearing this presentation at several Forage and Grassland Council meetings, I finally decided to challenge the Ph. D. presenting this new gospel. "What if you compared the sprayed pasture to a management-intensive grazed pasture?" I asked.

"Well, that was not part of this experiment," he responded. Of course it wasn't — it would have beat the sprayed pasture hands down in every measurable area: profit, productivity, animal performance. But the research routinely measures the latest crutch against some sick patient, never against a well patient.

It's like research showing that everyone ought to have a crutch because it allows people to walk better. The only problem is the research failed to bring in non-crippled people on the study. Comparing crippled people to crippled people, they found that crutches make people walk better. Conclusion: All people should use a crutch. How ridiculous can you get?

And yet research does this all the time. "We've compared this flawed model with this flawed model, and we've concluded that product X is beneficial for ***all*** models." It never occurs to them to check against unflawed models. This type of research is so pervasive in our culture that we would do well to question every single piece of advice coming from conventional channels. It's amazing how creative eggheads can be at spouting misinformation. Unfortunately, linear scientific methodology allows for only one variable at a time, and too much research is operating under the confines of this limitation.

Anytime a claim is made that a product will boost yield by X percent it is important to ask: "Compared to what?" The alternative agriculture community uses this same system, when declaring that such and such a product produced amazing results: "Compared to what?" Generally the comparisons are ***not*** to compost, earthworm castings, non-tillage perennial polyculture systems or keyline watering systems.

Research that shows the efficacy of wormers doesn't address controlled grazing, diatomaceous earth, herbal remedies or Shaklee Basic H. Research touting the efficacy of ionophore implants doesn't take into account cytokinins in kelp, nutritional differences in the meat or health problems associated with the increasing use of hormones. It doesn't take into account whether or not these products increase an animal's susceptibility to worms or pneumonia. I'll not belabor this point further, but I hope I have made a convincing argument for all of us to exercise skepticism over current research findings and their applicability to our situation, the comprehensiveness of this research (i.e. are there not other efficacious non-tested alternatives?) and the need for further equally flawed research in the future.

To get us back on track, then, the third thing we cull for is resistance to parasites and disease. We know we can do this culling accurately because we don't interfere with the animal's natural tendencies. All our animals are equally uncrutched; the cripples show up quicker and more accurately that way.

Fourth, we cull for beauty. Now you know I've fallen off the deep end. What kind of esoteric, unquantifiable category is this? Every herd of cows has what I call the spotlight cow and the embarrassment cow.

Over the years, we've found a direct relation between beef

eating qualities — taste, tenderness, dress-out — and the beauty of the live animal. We haven't used taste panels. We haven't used sophisticated tissue-density measurements. But we've listened to customers, which include gourmet chefs. And without a doubt, we've found that the spotlight animals also dress out the best and put the most palatable meat on the table.

A spotlight animal is the one that can walk through a mudhole and come out looking ready for the county fair. This is the animal that shines up first in the spring, that sheds that winter shag earlier than the rest of the herd. This is the one that stays fleshy when others may look a little thin in the hips. This cow's tail stays squeaky clean, and never carries telltale signs of manure.

She doesn't go and pant under a shade tree, but grazes longer than the rest. She doesn't hang around the water trough, but goes out on the hill and lies down. If you go out late in the evening and sit down in the paddock where the cows are grazing, she's the one that comes up and licks your ears and nuzzles your britches. When the flies are buzzing on a hot, sultry day, she's the one with the clean face, appearing as contented and placid as if it had just frosted last night. She's the one you want to point out to all the visitors, and say: "See number 5, there? Lord willing, in a few years they'll all look like her."

Lest I be misunderstood, let me articulate some things she is ***not***. She is not necessarily the biggest cow. In fact, our biggest cow at any one time has never fit this description. We're not talking about size. We're also not talking about the boss cow. Every herd has its echelon of leaders and followers. The boss cow, the top of the "peck order," may or may not be this one.

By the same token, she's not necessarily the lead cow. Controlled graziers know all about lead cows: they're the ones who lead the herd through the gate to the next paddock. They're the ones who first see you coming and know a paddock shift is imminent.

She generally is not the fattest cow during lactation. If she is, she's probably not milking as heavy as she should and will consequently wean off a small calf. But even when she's working hard during lactation and breed-back, she's clean and shiny; just a good working animal.

The bottom line is that in this consideration you're culling for animals that fit your operation. They metabolize your forage well. They don't tear up gates or fences. They are team players, and they're your stars.

The fifth and final criteria is calf size. Too often this aspect is shoved way up high on the criteria echelon. But for us this is last. I'd rather have a truckload of healthy 900 pounders than a bunch of unhealthy 1,000-pounders propped up with several visits through the headgate and $10 per head of pharmaceuticals. Even if the crutches pay for themselves, the hassle to the animals and to me of headgating them to administer the crutches, the deterioration and marketability decline in the meat, and the masking of genetic propensities keep such expenditures from being a bargain.

But all things being equal, we like a nice sized calf. Generally if we adhere strictly to the unassisted birth, beauty, and fertility qualities listed above, size takes care of itself. It's almost like nature maintains a proper functional size and weeds the excessively-sized animals out through maternal, fertility or beauty qualities. We do not mind having different-sized calves because some customers like larger ones and some want smaller ones.

If all we had were cookie-cutter sameness, where would the variety be to meet all the variables for our patrons? Besides, one of the centerpieces of natural farming is biodiversity.

Let's preserve this diversity within our herds. And this brings us full circle to the original theme of this chapter: individualized animals. What could be more diverse than a million small herds uniquely chosen to fit each producer's niche of God's creation? That would be the ultimate diversity. Let's be committed to finding what "works for us," and let that be the underpinnings of our "cow type."

Chapter 14

Choosing a Bull

I believe in compatible genetics. What good do Equivalency Progeny Data (EPDs) from a production model devoted to grain, pharmaceuticals and ionophores do for a producer of salad bar beef? Nothing.

It's like studying orange tree data when you're planting apple trees.

In our stupider years, we've used the top gaining bull at the test station. We've used artificial insemination. Both were major flops. The genetic criteria used to select those bulls were completely different than the genes we wanted. What does it matter to us if a bull can inhale and metabolize more grain than his buddies? There is absolutely no equivalency between performance on grain and performance on grass.

In fact, there's an old saying around here: "A bucket of grain can cover up a lot of weaknesses." It's common for a hotshot bull to fall apart when he's turned out on grass.

As I see it, grain testing is done for two reasons. First, it does make the bull grow faster and that allows the seedstock producer to display bigger numbers. It's back to the old "bigger is better" mentality. Second, since most beef ends up on grain, it does measure an animal's ability to perform on that diet. Grain-performing bulls will usually produce grain-performing heifers and steers.

But for salad bar beef, these goals are inappropriate. In fact, they can work to our disadvantage. For example, when we've used these kinds of bulls, we've had to pull calves — lots of them. One year we pulled all but one calf out of a heifer. I know all about hothouse bulls. I'll take one live 60-pound calf to a truckload of dead 100-pound calves any day.

Poor mothering instincts, skeletal problems. You name it, we had it. In fact, only one heifer from the artificial insemination trial lasted more than one lactation in the herd. We were asking these animals to perform in a model that was completely foreign to them, and they simply couldn't do it.

So how do we shop for a bull? First, we want a purebred. I didn't say "registered." That's not necessary, but I do want a purebred. Since we use crossed cows we want purebred bulls. The more we hybridize, the more slippery the genetic soundness. When we get farther and farther away from purebred genes, we lose semen potency and who knows what else. Nature maintains purity through natural sterility. Philosophically, I have a problem with crossbreeding and crossbreeding until we don't have any pure seedstock left, regardless of whether it's animals or plants.

Second, we look for a seedstock producer that is as close to a salad bar beef producer as possible. We don't want to see grain bins and bankruptcy tubes (silos) sitting around. We don't want to see a refrigerator full of medications and vaccines and we sure don't want to see a bunch of lick tanks and ionophore-enhanced mineral blocks.

All of those crutches, or shortcuts, reduce the likelihood of salad bar performance. Furthermore, use of these materials masks genetic propensities. How do we know how parasite-resistant a herd is when it's constantly doped up on Ivomectrin? How do we know if the herd has genetic resistance to pinkeye when they're vaccinated and medicated throughout the season to keep them from getting it? In fact, how do we know what kind of immune system we are getting if it is constantly suppressed with medications and pharmaceuticals? The only way to know what genetic material we have is to allow both strengths and weaknesses unadulterated expression. Anything else leaves us wondering if our animals look good because they are good or because we can afford pharmaceutical tricks.

We want to see a controlled grazing program so that the bull knows all about aggressive grazing techniques, electric fence and routine paddock shifts. We'd like to see a mineral program centered around kelp.

Third, we look for proven progeny. I could hardly care less about EPDs, but if I can see some calves out of that bull, then I know something. I couldn't disagree more with the mentality in the industry that sends bulls four or five years old to hamburger joints. I've heard Ph.D.'s say that genetic improvement is coming so fast that if you buy a four-year-old bull, you're already two years behind the newest improvements. Hogwash. Those guys must be getting kickbacks from seedstock producers.

I think the risk of unknown performance is greater than any loss we might incur with having two-year old genetics. Besides, anybody with any dirt under their fingernails knows such a statement is ludicrous. Only technology can change that fast; never nature.

From a self-protection standpoint, too, the older bull has merit. If a producer has kept a bull around for a few breeding seasons, that means he must have been satisfied.

If the bull throws calves too big to deliver naturally, or has some other defect, he doesn't last more than a season. And if he lasts more than two seasons, he must be producing healthy heifers and steers that grow well. I like to see progeny walking around.

An awfully lot of good old bulls of proven performance go into hamburgers every year while unproven young bucks enter the system and cause problems. One of the advantages in a salad bar beef program is that the nutritional level is high enough in the herd and the herd is bunched up close enough that a bull need not work nearly as hard to breed. Often bulls can breed 50 or more cows under a controlled grazing program when even 30 cows might be too many under a continuous grazing program. When the bull has to chase the cows all over creation, it takes a lot more energy than if they are within an acre or two at all times.

I know this notion may not sit well with seedstock producers, but good, proven bulls could gain in value rather than declining in value if more cow-calf folks would follow this advice. As certain seedstock producers gained a reputation for their bulls, the value would increase because if I can sell a bull for more than I paid for him, I'll spend more up front. There is no reason why bulls can't be used up to 10 years or more.

Much has been written about how to make the cow's life easier; perhaps it's time to put some attention on making the bull's life easier. A mature bull does not need the level of nutrition that a young bull does. Consequently, if he gets an extremely high level, like a salad bar, he can put more energy into breeding. But if he has to walk a couple miles a day looking for a morsel to eat, he has less energy left to make semen and breed. What is good for the cow is good for the bull.

In short, when choosing a bull we look for a purebred as close to our geographical area as possible that's been performing under as similar conditions as possible.

Chapter 15

Handling Facilities

One of the drawbacks with large animals is that we can't just walk up to them and dress a wound or help them deliver off-spring. That is one of the advantages of small stock like rabbits and chickens.

While this is by no means a definitive analysis of livestock handling facilities, it will offer some basic guidelines. As usual, there are as many different corrals as their are farmers. Certainly I have never built expensive or elaborate corrals. I am a functional person. Strictly utility. If it works, I don't care if the angle is a couple degrees off right.

Ideally, a corral should be located so that on some pasture shifts, running the herd through will be an easy thing to do. Several times a year we go through the corral as part of the normal pasture shift. Only a couple of times a year do we actually do anything in the corral. The cows think that going into the corral is just part of the pasture shift and this takes away the stigma of going in. I can literally call them from the paddock, walk ahead of them down the lane, step aside at the corral gate, wait for the last one to go in, and then close the gate on them.

"Getting the cows in" is a simple matter. If the corral is located somewhere near the middle of the farm, that will facilitate this routine procedure and make it much simpler to get the cows in when you really need to. The only times we do anything to the cattle when they enter a corral are when it is time to castrate calves, pregnancy check cows, sort to sell and sort to wean. The only time we send them through a headgate is to castrate and pregnancy check. A two-year-old heifer will have never been through a headgate. We do not pregnancy check heifers until their second calf.

From the initial corral that is large enough to hold the whole herd, you need smaller corrals into which you can sort smaller groups. It's always easier to take the many from the few than the few from the many. The older cows are generally easier to sort away from the calves than the calves away from the cows. The cows are smarter and they want to break back to the cutting gate quicker than their calves. They understand the escape pattern quicker than the calves.

Again, Bud Williams and Burt Smith are the gurus of cattle handling; essentially they point out that every animal has a flight zone. As we put pressure on the group of animals, they want to break back past us if they can't get away from us the other way. If we have them bottled up in a corral, and begin walking toward them, they will eventually turn around to face us because they feel threatened, and then if we keep walking they will charge past us to break away.

They are going to go somewhere to relieve the pressure as we "invade their space." If we will position ourselves to take advantage of the flow of animals — one of the strongest inertias in the world — we can let animals break past us or turn them back into the mob by slight backward and forward motions. Think of yourself as a little finger, moving in and out of the traffic flow — but ever so slightly.

I recommend at least three corrals, interconnected, into which you can sort. Obviously you can use a lane with sorting gates and lots of people, but I do my sorting by myself or with my son, Daniel, so we generally sort from the big group into one of the smaller corrals that opens directly into the main corral. As we get down to the "keepers" we can put them in another small corral. If we need to pull off some more "non-keepers" we can move the big group in the small "sorting into" corral over to the other small empty one (we're

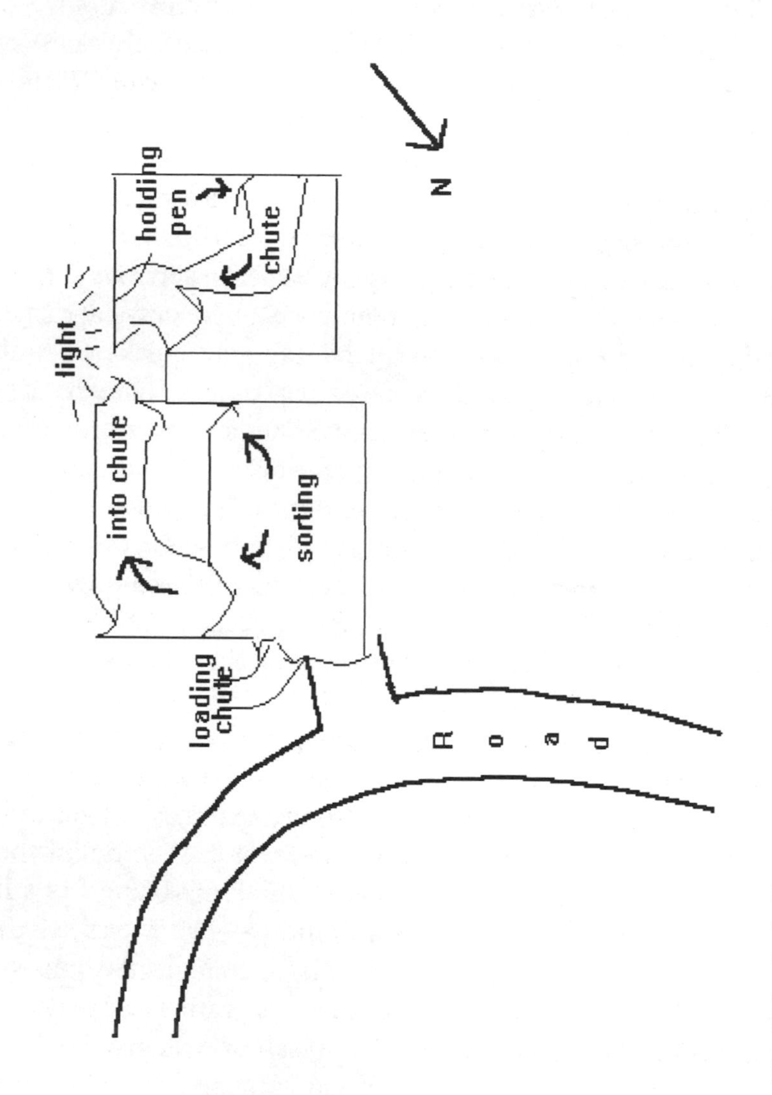

Fig. 15-1. *Cattle like to move in a circular pattern, toward light, and go out where they came in.*

just sorting two groups) and continue bleeding off the many from the few until we're down to just "keepers." Then we can amalgamate the "keepers," run the "non-keepers" back into the big herd corral, and move the keepers into the holding pen for the headgate.

Cattle always want to come out where they went in, they want to go toward light rather than dark, and they do not like walking into direct sunlight. They like to move in a 'J' pattern, or circle back around so they can exit where they entered.

We have a holding pen big enough for half a dozen cows. We can open a nice wide four-foot gate at the end of it to let the cow go in. The alley narrows down as it makes a 'U,' sending the cow straight for the spot where she entered the pen. In other words, to enter the holding pen, the cows have to cross in front of the headgate. As they enter the alley, they think they are exiting.

Our holding pen and headgate are in an awning in the barn. The light comes in from the eastern end. As the cows make the turn in the alleyway, they see light beyond the headgate and this draws them toward it. I built a simple gate out of pipe and angle iron. A rope attaches to the back of the alleyway. When the cow goes in, I close the gate behind her and grab the headgate rope. When she puts her head through I give a tug and dogs on the moveable catch grab notches, which lock it in place. No noise and very little expense.

Four two-foot gates in the alley allow access to any quarter on the cow. We used tall poles, heavily cross-braced on top, to give us a simple but functional system. I do not like noisy headgates. By the time the cattle have listened to that for awhile they think they'll be killed when they get in there. I like to work them quietly and safely. Certainly there are times when we have a belligerent animal. But that should not be a problem that happens routinely when we work the livestock.

It would behoove all of us to pray for patience in this regard, especially those of us who are aggressive Type A guys busy about our little entrepreneurial salad bar beef enterprises.

Handling facilities should be simple, but they need to be strong and functional. Again, do not go out and buy the latest multi-hundred dollar gizmo just because it has shiny paint on it. Something a little quieter and more flexible can often work just as well, if not better.

Chapter 16

Electric Fencing

Whole volumes have been written about building electric fence. Far be it from me to think I could offer a comprehensive guide in one short chapter — or even a short book. But, in keeping with the other ideas I've espoused, I do intend to tell our story and offer a few tidbits along the way.

The important thing to remember about fencing is that it is either a physical or psychological barrier. A physical barrier must withstand the abuse of 1000+-pound animals bumping into, rubbing against, and occasionally directly assailing it.

But a psychological barrier is completely different. The animals do not touch it. All they need to do is see and remember. If they can't see it, and/or if their experiences with it have not been memorable, it may as well not exist. If, however, they can see it and the memory of their last experience is indelibly implanted in their heads, they will respect it as much as a concrete barricade.

Before setting up the first foot of electric fence, then, commit yourself to consistency. Promise yourself you will check it every day. I don't know how many people have complained to me that they can't get their cows to respect their electric fence. I go over and there is no spark in it, or some of it is covered up with weeds so the animals can't see it, or some of it is down to a couple inches off the ground. Electric fence is ***not*** something you put up today and walk away from, like Cadillac quality Red Brand six-inch field fence.

That kind of fence can be built and left alone for 40 years if the posts don't rot. But electric fence is completely different. If you are used to no-maintenance fence, understand that electric fence must be checked virtually ***every day***. I am belaboring this point because so many folks think electric fence takes care of itself. Let me assure you, there's no free lunch. The fact that you can completely fence a 100-acre farm for a few hundred dollars should not mislead you into thinking that fence will be as self-maintaining as the one you would build for thousands of dollars. It will not. It can be cheap, versatile and require monitoring, or it can be expensive, inflexible, and low maintenance.

With that in mind, let's talk about electric fence. We have a little Craftsman fence charger, 1963 model, that still energizes all our fence and just keeps plugging away. Each year a fuse blows out in a lightning storm, but that's all that has ever gone wrong with it. It was one of the first solid state chargers to come out. In fact, it is so

Photo 16-1. *Functional, poor-boy electric fencing. Porcelain white knob insulators, 9-gauge aluminum electric fence wire, homemade locust gate handles, lag screws bent in an 'L' to hold gate knob, and a 5-foot locust stake pushed into the ground with tractor front-end loader. No brace posts.*

ancient that written in big, bold letters across the front is: "Solid state. No moving parts."

My point here is that multi-hundred dollar units are ***only*** appropriate when you are energizing miles of multi-wire fencing or for sheep and goats. These little $75 units and less, that sell at the local farm store, are more than adequate for most applications. You can buy a Cadillac or a Ford Escort; both will get you where you want to go: one just does it in more style. Electric fence is like that. You can spend yourself into bankruptcy buying everything first class. Get something in the middle of the road and you'll probably be fine.

Make sure you have a good ground. All the installation pamphlets from electric fence companies are correct on this point. Remember that the ground is what receives the electrons (the minus charge) as it comes back to the charger. The spark in the wire is positive. The reason you get shocked when you touch it is because you complete the circuit between the minus (ground) and plus (fence). If these electrons do not have a good sending-receiving unit (the ground rod) the completion of the circuit will be poor, kind of like corroded terminals on a car battery.

Make sure that ground rod goes way down in the soil where it will never get dry. Ideally, the ground should be hooked to a culvert or water pipe.

The next important principle is to put a lot of cutouts in the system. Again, Jim Gerrish had a standing bet with the fellows at the Linneus, MO experiment station that if he could not trace down a short in 30 minutes or less, he'd treat them to dinner (I think it was a non-salad bar hamburger). With a series of cutouts, a short can be traced down systematically in a few minutes. You should be able to isolate the energizer from the fence. Then the fence should come to the energizer in at least two or more trunk lines.

Visualize a water system. You have a pump, big trunk lines, smaller lines, and then individual lines into homes.

An electric fence is much the same. The more valves you can turn off, the quicker you can isolate a problem. A short in the fence is like a leak. You want to be able to start at the charger, isolate which half or third of the whole system the problem is in, then go to the next cutout, and so on until you find the leak. By the same token, you want to be able, if you're on the back forty with the cows and have no spark, to disconnect each piece as you work your way back toward the energizer. Most of the time you will find the short in the far appendage of the grid. The key to quick discovery of a short, obviously, is to have the capability of isolating chunks of the system. This requires many, many cutouts: the more the better.

These cutouts can be sophisticated switches, little alligator clamps from Radio Shack or other electronic supplies stores, or a hooked wire. I know some people will cringe when I say this, but we use high tensile aluminum (Tipper Tie) wire and it makes such good contact, we just make loops and let them pull against each other. The aluminum high tensile is springy so it holds good contact, as opposed to regular smooth steel wire which has no tendency to spring.

Every tractor and every truck on our farm has a pair of insulated pliers. These are our tool. With these I can hook up and unhook these little cutout loops. Seldom am I without or farther than a few yards from a pair of insulated pliers.

Permanent fence, for us, consists of 60-inch locust posts big enough to hold a nail. During wet spring and winter days I split up these short posts with sledge hammer and wedges into stakes big enough to hold a nail. In the early spring, after the thaw and while the ground is soft, we sharpen one end of these stakes with a chainsaw and push them in the ground with the front end loader. Setting them on 15-yard centers we can put up a mile of fence in a hurry.

Any rot-resistant locally-available wood will work for the posts. They need to stick out of the ground far enough to hold an insulator at animal-nose height. Cattle will go over a wire long before they'll go under a wire. It's better to have the wire a little high than a little low. Besides, a high wire encourages under-fence grazing, which reduces brush and weeds under the wire.

We like the aluminum wire because it does not require brace posts. It is light enough that a 100-pound pull makes it good and taut, assuming the distance is not more than several hundred yards.

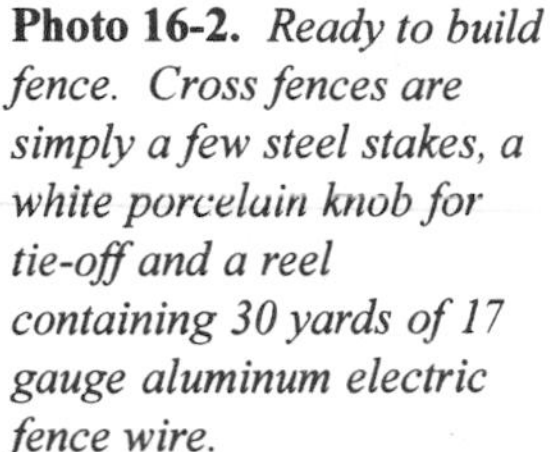

Photo 16-2. *Ready to build fence. Cross fences are simply a few steel stakes, a white porcelain knob for tie-off and a reel containing 30 yards of 17 gauge aluminum electric fence wire.*

We use 9 gauge for permanent wire and 17 gauge for temporary. We virtually eliminated deer damage when we shifted from steel wire to aluminum — apparently because they could see it better at night. The aluminum glistens in the moonlight. Deer would rather go under than over a fence, so running the wire a little on the high side also helps eliminate deer damage. If they can duck under it easily, you'll scarcely ever seen one go through it.

The aluminum conducts much better than steel and never rust. It is more expensive as an initial investment, but once purchased it seems to last forever. We attach it to ends with little porcelain spools, and although I have read that these tend to crack, we have plenty that have been up for more than 30 years and they appear to be as good today as they were the day they were installed.

We make our own reels out of discarded self-feeding stainless steel welding wire spools. These normally end up in landfills. We have a local industrial shop that saves them for us. About once a quarter we go down and give the manager a few dozen eggs. They go out of their way to save the spools for us and the eggs sweeten the pot. We bore two half-inch holes through the webbing and put 11-inch carriage bolts through for handles. Using an acetylene torch, we pop a hole through used hay mower knife sections to act as a nice big washer against the spool edges and webs.

We end up with a spool that will easily hold a quarter mile of 17 gauge wire, with two handles at just the right position for fast reeling. It is not as snazzy as what's available commercially, but it's a lot cheaper and very functional. I can roll or unroll fence as fast as I can walk, and because these are so cheap we have a pile of them. If

Photo 16-3. *Homemade locust rounds serve as excellent handles. Getting ready to run a cross-fence: hook handle into piece of insulative twine to keep spark out until ready to energize.*

one cracks, we just make a new one. They last for years. Having so many allows us to lay out many cross fences at a time. That way if we are going to be gone for a few days, or if we get real busy, we can lay out a bunch of paddocks ahead of time and go for several days without changing any fences.

We make our own gate handles out of locust rounds. We just drill a 3/8-inch hole through the 6-8 inch piece (2-3 inches in diameter) and put a heavy 9-gauge steel wire through it, bending it for a hook and putting a loop on the back end. Only a couple times a year does it get so damp that the handle will conduct some current. Otherwise, it's very functional. If you run over it with a tractor tire, it just smashes in the ground but will not break..

All of our cross-fence reels have a gate handle attached to hook onto the permanent fence. We use baling twine as an insulator to let us hook up the handle to the hot fence and run the cross fence out. Just a little 2-inch loop of twine is plenty to insulate the reel

Photo 16-4. *Reeling out -- or up -- electric fence using homemade reels made from self-feeding stainless steel wire spools. Two 14" x ½" lag bolts and four junk sickle bar mower knife sections with holes burned in (to spread across plastic webs, as washers) offer practically cost-free reels.*

Photo 16-5. *At the opposite end, tying off to internal permanent wire. Pull fence up to porcelain knob, tie baler twine through and pull tight around post; hang reel from permanent fence with baling twine.*

from the hot fence and allow you to roll out the cross fence. At the other end, we use a baling twine looped through a porcelain knob, pull the wire through and give it a couple hard 90-degree twists, then hang the reel off the other permanent fence with a piece of baling twine. Twine is wonderful material because it's free (leftover from bales of hay we feed out) and it rots.

If we are pulling from an internal permanent fence to a boundary nonelectric fence like high tensile or regular woven wire, we use a little 'L' that we call a tie-off and a sock. The 'L' is welded to a piece of ½-inch pipe that slips over the electric fence stake. A white porcelain knob goes on the 'L' and allows us to pull the small-gauge cross fence. Another piece of 1/2-inch pipe with a smashed end and two little ears welded on provides spool storage. The reel is mounted horizontally on the sock. A baler twine goes around the stake to the non-electrified fence as a brace against the pull. We also tie a piece of twine around the reel as a safety against the wire popping off, unwinding and shorting out on the stake below. This procedure gives

Photo 16-6. *At nonelectric boundary fence or freestanding end in a field: reel stores on sock, fence pulls to tie-off, and twine around reel keeps from unrolling. All homemade, but extremely functional -- and cheap.*

enormous flexibility. It can also be used out in a field with a simple brace stake and twine if we want to make a corner out of two temporary wires, for example.

Again, I am just trying to illustrate how simple an electric fencing system can be. It need not be intimidating or confusing. There are plenty of knowledgeable sales folks, plenty of reputable companies out there to advise you. Just don't buy everything hook, line and sinker. We purchase 3/8 inch rebar 42 or 48 inches long for as little as 50 cents or a dollar apiece as portable stakes. Yes, they are heavier than fiberglass tread-ins, but they are much cheaper and they never break. They never crack when it's cold. Most of the year

Photo 16-7. *Freestanding end in field. Note brace stake. Electric fence allows complete flexibility of paddock size and design.*

they can be pushed in by hand. The more fertile your ground, the easier they push.

Some people weld 4-inch pieces to these stakes so they can be pushed in with your foot. A quiver, too, to carry them in to leave your hands free is certainly a good additional refinement.

Just keep the system simple and think function, function, function. You will learn as you go along, so do not overbuy at the outset. You can always buy things you need as you refine or as you can't figure out an alternative. But once you've bought, you can't get rid of it.

Chapter 17

Paddock Layout

A controlled grazing setup is a combination of permanent electric fence, portable electric fence, water, forage, animals and management. We will devote our attention here to the layout of the fences.

The first consideration is to lay the permanent fences on the keylines. An Australian named P. A. Yeomans invented the term "keyline" and wrote several books about the concept and its adaptations. It provides the basis for his book, *Water for Every Farm*, which postulates that the fertility weak link is enough water to grow enough vegetation to feed the decomposition process in the soil. But the foundation point is the keyline. *[See Photo 9-1, page 48.]*

The keyline is the point where a ridge breaks over to a hillside and the hillside breaks to a gentle valley or terrace. While it is certainly not as profound in flat country, in gently rolling areas it is quite profound. A ridge will roll off toward the hillside and then there will be a fairly sharp break point — just a couple of feet — where the hillside really breaks away from the ridgetop.

By the same token, at the bottom of a hillside, a rather sharp break occurs as the ground flattens out into a swale. These break points are called the keylines. It is a concept that is obvious once we're exposed to it. When I first read about it, I took a walk and sure enough had no trouble noticing the break points. The keyline is important in paddock layout because the big differences in microclimate occur before and after this point.

The rule of thumb is to have as similar an area as possible enclosed in one paddock. Hydrology, aspect, soil type/depth and vegetation type are highly variable depending on the topography. For example, the ridge top will be fairly dry, which means it will be drought-prone in the middle of the summer. By the same token, the ridgetop will be dry and early growing in the early spring. The ground will accept cattle earlier (without danger of pugging) than a nearby swale. Pugging results when cattle tromp the soil during damp conditions, destroying the existing sod, pressing out oxygen, damaging soil structure.

The south-facing slope will tend to be hotter even than the ridgetop, greening up earlier in the spring and staying green longer in the fall. The north-facing slope will be cooler, damper, have deeper soil, retarded spring green-up and be more prone to pugging damage. But it will stay green during a drought because of the cooler, more moist conditions. Not receiving direct sunlight, the north slope will generally grow less forage volume than a south slope will, but it will grow it during those critical dry summer months when the south slopes may bake brown.

Then the valley, or swale, has its own characteristics. It tends to be more moist than the north slope, which is wonderful in the middle of the summer, but devastating during "mud" season. It tends to have the deepest soils because it collects the sloughing from the hillsides.

The point of all this discussion is that if we graze a south-facing slope, a deep swale, a ridgetop and a north-facing slope all in the same paddock, our mismatching will reduce the efficiency of the grazing program. For example, if we put the cows in during a drought when the north-facing slope is ready to graze, chances are the south slope is baked brown, has not regrown to energy equilibrium, and we will overgraze that area, damaging it. By the same token, if we turn the cows in when the south slope first greens up in the spring and the cows walk down in the swale, where the soil is still cold and wet from winter and the grass is dormant, we damage the sward through pugging and muddy up the cattle.

The keyline at the bottom of a hill is often damp. Water ducts out from the hillside and eases into the swale at that point. We need to be careful about placing a fence above the keyline so that the cows walk right in that seep area. We want to come down far enough that we can get the advantage of the seep without muddying it.

All of this lends import to the admonition Bill Mollison, founder of Permaculture, makes repeatedly: after gaining access to a piece of ground, do not do anything for a year. Walk it once a week, making notes about where the wet spots are, where the warm air flows, where the cold air flows, where the dry spots are and the type of vegetation in different areas. A spot that grows wild grapes, for example, will probably grow domestic grapes exceptionally well. A spot that the deer keep grazed off is apparently able to grow especially succulent forage — mark that spot for a weaning area, for example.

The land, with both its assets and liabilities, should be the foundation of what we do. Too often we force it with square fields and inappropriate production models. Wes Jackson's wonderful book, *Meeting the Expectations of the Land*, explains this point eloquently. The world is full of orchards planted in the wrong place, nurseries planted in drought-prone or frost-prone areas, steep northern hillsides planted to grass, and the like. As soon as we try to force on the land something it is not suited for, we either have to use capital-intensive mechanisms (like irrigation and field sprayers) to maintain production, or we reduce conversion of solar energy and take the loss.

For example, deciduous trees are about 50 percent more efficient at converting solar energy into biomass than are forages. In many areas, including ours, north slopes should be left in trees and south slopes planted to forages. The southern aspect tends to be drier, which is not as conducive to good tree growth, and the northern aspect gets more indirect sunlight, is more efficiently captured by deciduous trees. This is why conifers in the eastern hardwood forests of the U. S. grow on south slopes. They are shade intolerant and can grow in hard, dry soils. North slopes, on the other hand, grow the premium quality hardwoods. The soil is deeper, the ground more moist and soft, and the trees are shade tolerant: they can thrive on much less sunlight.

Eastern aspects are more like northern slopes than southern, and western aspects are more like southern than northern. The morning sun is never as hot as afternoon sun. This may seem like wandering off course for a book about salad bar beef, but these factors play a critical role in paddock design and layout. Mismatching the grazing to the land keeps us operating at half throttle instead of full throttle. If we do not fully capitalize on the assets or shepherd the liabilities, we cannot convert as much solar energy into plants. The less solar energy we metabolize on our farms and ranches, the less productive they will be.

Straight fences are not found in nature. Placing permanent fences on the keylines will not necessarily insure that we always correctly match the grazing to the land, but it will go a long way toward reducing abuse. Animals tend to walk on the contour, not up and down. While keylines and contour lines are not the same, they are similar enough to capture this tendency of animals. The animals benefit, the land benefits, and the bank account benefits when we place the permanent fences on the keyline.

The permanent fences, which can be just a single strand of wire, define the permanent pastures. They also define what is ***not*** in pasture. For example, a fence should run around ponds, along riparian areas, and along woodland. Much like raised beds in a garden identify the cultivation area and hence help us focus our labor and creative energies, so identifying what is pasture and what is not helps focus our efforts on what is important. We need not worry about that weedy little pond overflow — it's not in the pasture. We need not worry about that dead tree in the woodlot — it's not in the pasture.

Certainly we may want to do something about these things, but they do not enter the grazing picture. At the risk of sounding like I'm advocating compartmentalizing the farm and not treating it as a whole, I do contend that we are more efficient at directing our attention toward single areas at a time. That is not to say we do not integrate and/or think about how to integrate. But when the cows are lounging in the pond, the aggravation of that problem will detract from our efficient paddock movement. The world's most productive and beautiful gardens are always clearly bordered — related to "ordered" — and our fields can be viewed as our gardens. This is true husbandry.

Novices would do well to put in as little permanent fence as possible. Use portable first, and whichever parts do not get moved in three years are the ones that should be made permanent. This way you know the permanent fences you put in have precedent because they work for you. It is better to put in minimal permanent and gradually switch it over than to put in too much permanent and try to make it work because you have too much labor and money invested in it to tear it out. Let the learning curve happen and stay loose.

With a network of permanent fence in place, look at the fields and see how long the temporary break fences will be. Jim Gerrish at the University of Missouri has done some outstanding work in this area and I highly recommend his research regarding building fence and laying out paddocks. He says that 200-300 yards is about as long as you want to run temporary paddock breaks. I agree.

If you have a field, for example, that would require a 400-yard temporary break, it probably needs a lane up the middle of it. The reason is twofold. First, the break fence is too long to be efficiently set up and taken down routinely. The paddock "slices" will probably be too thin, resulting in long, narrow paddocks. The more square the paddock, the more efficiently and evenly the stock will graze. The longer and thinner the paddock, the more steps they will take per mouthful of grass. Whatever the cow tramples, she won't eat. Minimizing trampling damage is key to increasing utilization. We'll talk more about that later.

A lane will allow you to shorten the breaks and square up the paddocks. Secondly, the lane allows more grazing flexibility. The key word to controlled grazing is "flexibility." Allan Nation has aptly pointed out that until portable electric fencing, grazing was like running a combine without a steering wheel or brakes. Portable electric fences are the brakes and steering wheel of the four-legged grass harvesting machine. The more flexibility we can put into the system, the better we can drive the cow.

A lane, with all the paddocks opening onto it, does not take much land but offers immense flexibility. We can always get the animals back to the corral without running back over grazed forage or running over ungrazed forage. If we are moving from paddock to paddock and suddenly realize we need to skip a few paddocks to let them grow for hay, we can go into the lane and into whatever area we want. The greater flexibility also allows us to better match the cattle to the land. The longer the strip, the more apt we are to be hitting dry spots and wet spots in the same break. Reducing the length facilitates homogeneous soil type and forage quality.

Although this picture is far too simplistic, it may help illustrate the overall goal for paddock layout. Imagine a ladder. The outside supports are the permanent fences and the rungs are the cross fences. Without the outside supports, the ladder will not work; neither will it work without rungs. It takes both. Laying out a paddock system requires both permanent fences and temporary fences. If all we have are main supports, or permanent fences, the system is too bulky, too heavy, too inflexible. If all we have are rungs, we can't put them up and take them down fast enough to climb. We need both rigidity in the permanent and flexibility in the temporary, or portable.

Running the heavy supports on the keylines, and then connecting the keylines, at right angles, with the rungs, we have the beginnings of a functional steering wheel and brake on our four-legged mowing machine. Paddock layout needs to follow the variances of the land and be highly flexible. Custom-fit, farm-individualized designs, will ensure efficiency and function for many years.

Harvesting the Salad Bar

Chapter 18

Grazing Philosophy

Forage grows in an 'S' curve.

That means that after being sheared, either by a grazing animal or by a mowing machine, it begins growing slowly and then picks up speed before tapering off again as it reaches physiological maturity. When the plant is sheared, it prunes off roots to focus energy in the crown to send forth new shoots. A nurseryman prunes saplings before transplanting for the same reason: if all the roots and branches were left on the transplant, the tree would die trying to maintain all the appendages. Pruning takes the pressure off the plant and allows it to put forth new growth and adjust to the shock.

The pruned off grass roots add organic matter to the soil. As the new shoots draw on stored carbohydrates in the crown and roots to send forth new shoots, the plant's energy is temporarily depleted. The weakened plant returns to energy equilibrium only after the new growth has gotten long enough to capture excess solar energy and regrow root mass. The roots will always mirror the plant's top growth.

We call this "pulsing" the pasture. The idea of letting the plant grow several inches, then shearing it off, and allowing it to regrow to that several-inch level is completely counter to conventional grazing today. In typical grazing, the animals are left on one field for long periods of time and the palatable species are grazed excessively and the unpalatable species are avoided. This constant

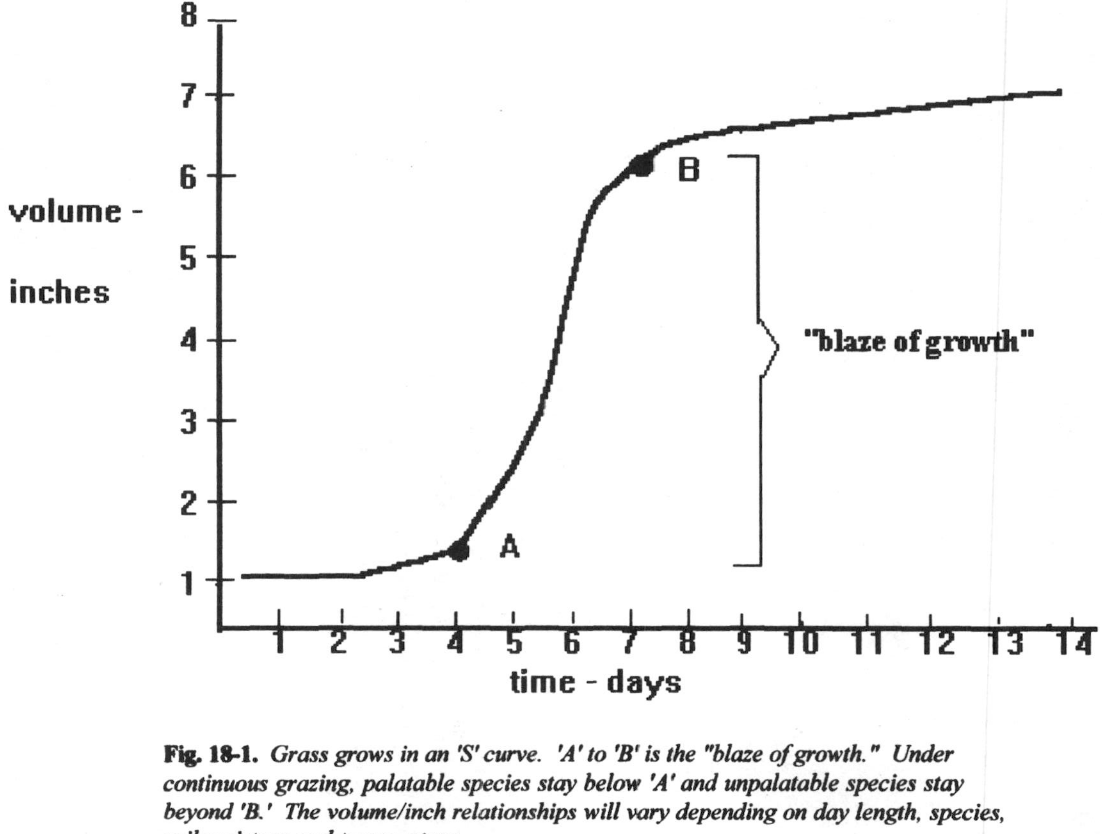

Fig. 18-1. *Grass grows in an 'S' curve. 'A' to 'B' is the "blaze of growth." Under continuous grazing, palatable species stay below 'A' and unpalatable species stay beyond 'B.' The volume/inch relationships will vary depending on day length, species, soil moisture and temperature.*

overgrazing of good species and undergrazing of poor ones produces a pasture sward that regresses to fewer and fewer good species while producing more and more poor species like weeds and brush.

Understanding the plant growth cycle is critical to maintaining productive forages. The fast growth part of the 'S' curve is called the "blaze of growth," and it is during this critical time that roots are regrowing and forage volume increases multiplicatively. At the second break point, or the top of the 'S,' the plant's energy reserves are restored; growth slows, and nutrient levels peak.

For optimum forage and livestock performance, then, the plants should be consistently harvested at the second 'S' curve break point: right after the blaze of growth and right before the maturity slowdown.

Unfortunately, most forages are harvested at the first break point: right after the slow growth period and ***before*** the blaze of growth. Under continuous grazing, palatable forage species are overgrazed while unpalatable ones are undergrazed. A continuously-grazed pasture always looks clumpy. The productivity is greatly affected because by harvesting at the first break point perhaps 20 times per season at a volume of 400 pounds per acre per shearing, we have a total of 8,000 pounds of grass harvested per season per acre.

But by allowing the grass to express itself through the blaze of growth and harvesting only 6 times at 3,000 pounds per acre, the total is 18,000 pounds of forage produced. The only way to achieve optimum productivity is to control when the plants get sheared.

At the same time, animals consistently grazing plants at the first break point will not get the level of nutrition that they should. Such short plants will have short roots, and therefore not bring up minerals buried lower in the soil. Furthermore, short-rooted plants are more susceptible to drought damage. Weakened plants will not jump back as fast after heat or drought.

This growth cycle cannot be charted on a calender. It fluctuates with the season and the forage species. Cool season forages grow faster during early spring and late fall. Warm season forages grow faster during the summer. Day length, soil moisture, and ambient temperature all bear on the rapidity of the growth cycle. Sometimes the cycle can occur in 10 days; other times, it may be 50 or in winter, 100 days.

The key is to never allow forages to be re-mowed or re-grazed until energy equilibrium is achieved. This is called "the law of the second bite." A rapidly-growing plant in spring, nipped twice at the first break point, will be weak for the rest of the season. In fact, many plants, especially the palatable ones, actually die due to continual weakening. An animal always eats dessert first. Given a choice between ice cream (clover) and the liver (ragweed) the cow will always choose the ice cream. That is why continuously grazed pastures, over time, tend toward weeds and woody species, and away from clovers and highly succulent grasses.

Just as a garden requires careful tending to maintain the desired plants and discourage the undesirables, so a pasture needs careful monitoring so that the shearing cycles will be in sync with the forage growth cycles. Both plants and animals benefit from such control.

Whether it is one goat in a half-acre lot or 1,000 cows on a 10,000-acre ranch, the principle remains the same. The animals should be moved from a grazing area (paddock) before they have an opportunity to re-graze the new shoots.

All we are trying to do is mimic on a domestic scale what herbivore populations the world over show us. Whether it is wildebeests on the Serengeti, caribou in Alaska, bison on the American range or reindeer in Lapland, multi-stomached herds are always moving onto fresh ground. They follow forage growth cycles and never stay in one place more than a couple of days. We call that short duration stays.

Secondly, they are always mobbed up. Predators keep them mobbed up. Even though they may have many square miles to graze, they are close together. This mobbing principle we call density. The two D's of grazing: duration and density. High density completely changes the herd's grazing behavior, its interaction with the soil, and its efficiency.

We achieve the same objective domestically with portable

electric fencing (our predator), allowing highly efficient 24-hour paddock shifts. In our county, the average cow-days *[see Chapter 21]* of pasture productivity per acre per year is 70. That means the average pasture acre will produce 70 cow-days worth of forage (Extension service computer models generally use 3-4 acres per cow as a baseline). We have had paddocks go up as high as 400 cow-days per acre in a season on this farm. The difference is tapping into the growth cycle. Everything in the system wins: plants, animals, and bank accounts. If for the price of some electric fence and some management time we can double our forage productivity, it's identical to doubling our land area for a few dollars. We've bought another farm for the price of some electric fence and better management.

The importance of monitoring the growth cycle and the benefits of doing so are well documented in Allan Savory's *Holistic Resource Management* and Voisin's *Grass Productivity.* All I am presenting here is the fundamental concept, nothing more.

Graziering is an art as well as a science. It is like any skill: no one can really do it until they actually do it. As Allan Nation says, it's like driving a car. You can sit in a classroom all day and learn the fine art of driving a car, but until you actually get behind the wheel and take it for a spin, you'll never really know what it's like or how to do it. Experience is definitely the best teacher. After making mistakes, you will begin to get a feel for it and it will become easier and easier. But the only person who can learn it for you is you. It is not a gadget you go out and buy. It is an art form that you mold and create to fit your own piece of ground.

The rewards are myriad. Watching the soil, forage and animals respond through your management rather than through purchased materials and gadgetry is truly fulfilling. I can't imagine why someone would want to go out and till and cultivate and worry about erosion and breathe soil dust when it can be so much greener and more enjoyable. Graziering is wonderful.

Chapter 19

Matching Seasons, Nutrition and Grass

Managing cattle, forage and soil is as much an art as a science. That is why people who think they can "buy" their way in are frustrated: it can't be done. Nature cycles with a certain rhythm, and this requires being in tune with the seasons and how these changes affect the whole salad bar beef enterprise.

When grass first begins growing in the spring, it is recovering from overwintering and needs to restore depleted root reserves. These tender spring plants could be viewed as "babies," and treated similarly. The more of a grass plant we shear, the greater the shock to the plant. Taking 70 percent of the plant is far more shocking than taking only 30 percent.

"Baby" grass plants cannot take heavy shocking. We reserve that for other seasons. At first grazing in the spring, then, we avoid high utilization of the forage, and graze lightly, utilizing only maybe 30 percent of the available forage. Distinct graze lines that are a trademark of controlled grazing should not exist early in the spring.

Furthermore, with the propensity for pugging damage early in the season, we want to offer large paddocks to the stock. In the first grazing cycle we try to cover the entire farm in 14-21 days, using huge paddocks and lightly "creaming" the grass. We add a few bales of junk hay to keep the animals' manure from getting too loose (runny) during this transition from hay to grass.

Photo 19-1. *Before grazing. Note length and thickness of pasture sward.*

Photo 19-2. *After grazing, only 24 hours later. Note evenness of graze, amount of residual (high) and height of sward. These variables change with season, type of stock, forage species and many other considerations. The foundation principle is close observation and monitoring.*

If we have some leftover stockpiled forage, this is a great place to start the spring grazing because the cows will seek out that old dead grass to balance their rumen. If all they get is what I call "candybar" grass, their manure will be overly runny and they can actually lose weight just like people do with a serious case of diarrhea.

Our goals during this first rotation are to:

- Minimize impaction to soft ground.

- Delay physiological maturity in all the fields.

- Stop feeding hay as early as possible.

- Minimize grazing stress (shock) to the grass.

The next rotation, we cannot keep up with the forage growth so we drop a few paddocks. In essence, this reduces the acreage in the grazing area as forage growth escalates in order to maintain a constant amount of forage. It's a plus and minus. The amount of forage required to feed the cows is constant; if the forage production is going up, the grazing pool of pasture must diminish to hold a constant supply. These extra paddocks can be cut for hay.

By the third rotation, the grass growth speeds up even more and additional paddocks must be dropped. Shortly after this grazing cycle, we're into June and growth begins to slow.

At the same time, the cow-equivalents are escalating. Baby calves are larger, the cows are lactating and beginning to cycle into heat, and the yearling calves are hitting full stride. We are faced with a declining growth in grass and an escalating feed requirement.

After mowing the dropped paddocks for hay, then, those areas are added back into the grazing acreage, or pool, to increase the area and again hold a constant supply of total forage produced. Because the ground is firm by now and the grass is stronger (more highly

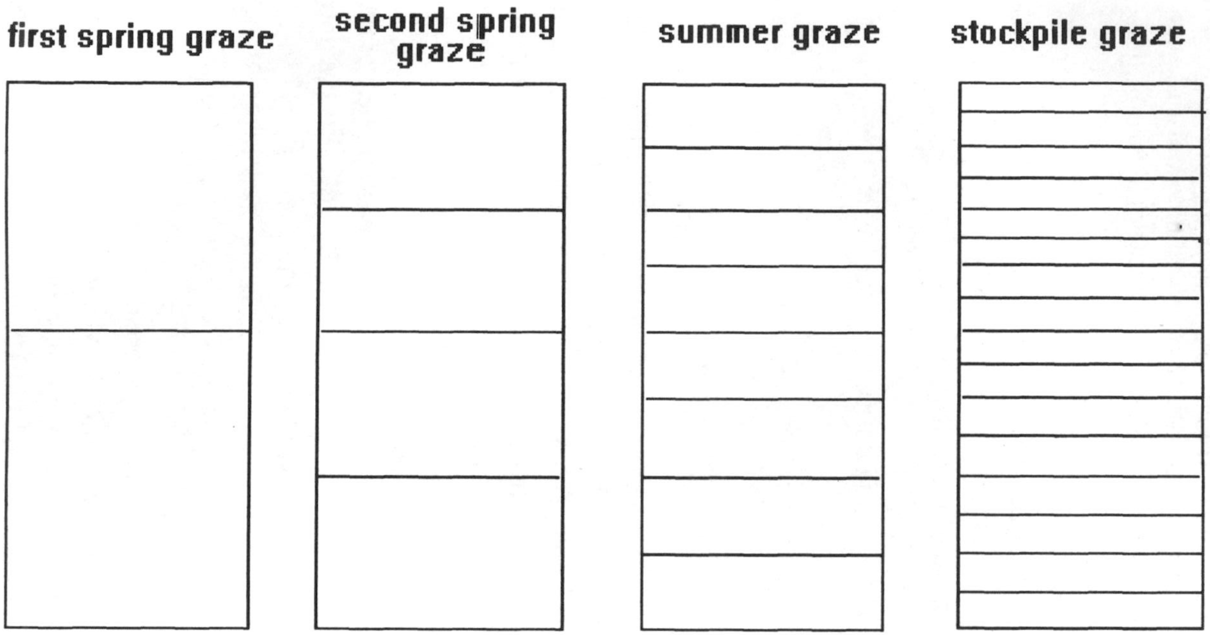

Fig. 19-1. *Varying paddock size to change the density, utilization and rest period is accomplished with temporary electric fence subdivisions. Paddock size -- number of subdivisions -- will change for each grazing.*

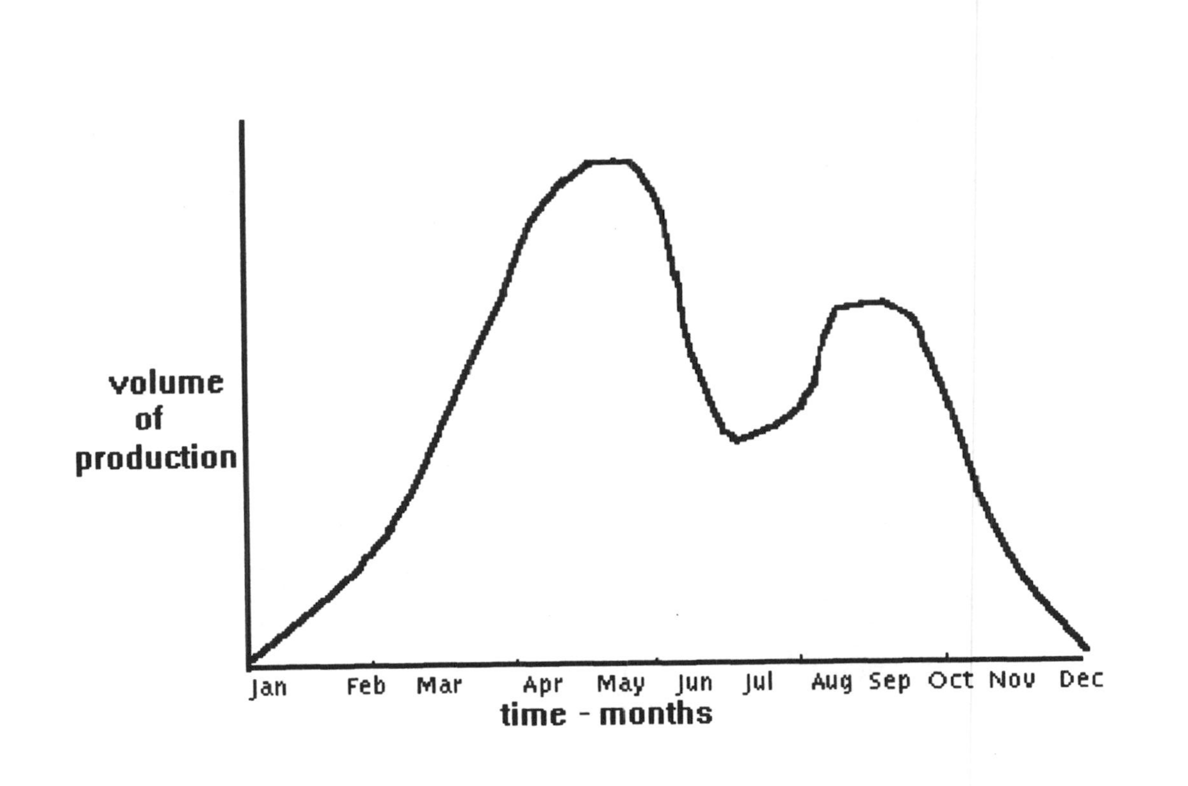

Fig. 19-2. *Nearly every forage species undergoes this growth pattern. The relationship between time and volume will vary depending on climate, species, fertility and season. Forage growth fluctuates throughout the year.*

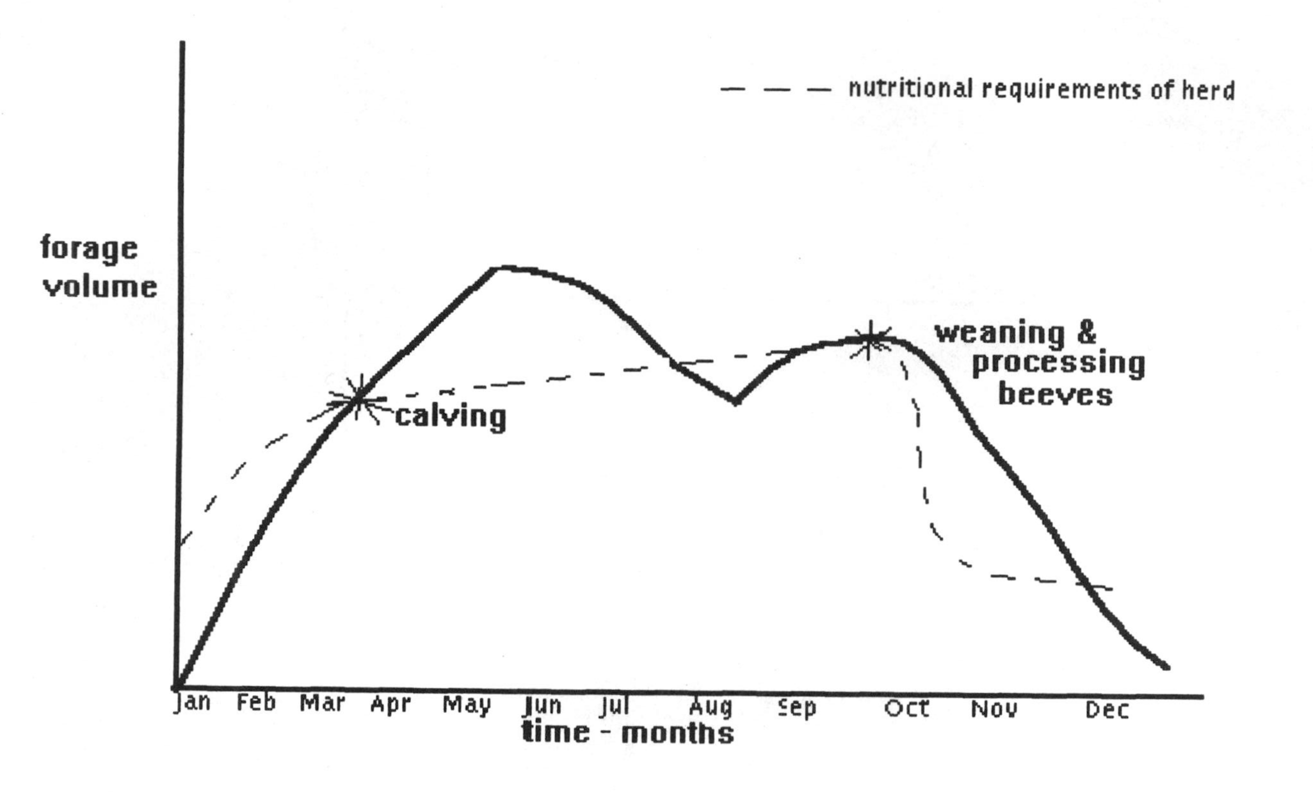

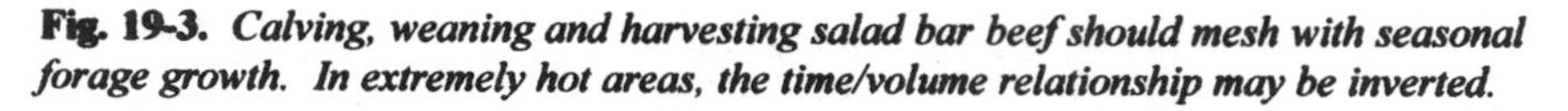
Fig. 19-3. *Calving, weaning and harvesting salad bar beef should mesh with seasonal forage growth. In extremely hot areas, the time/volume relationship may be inverted.*

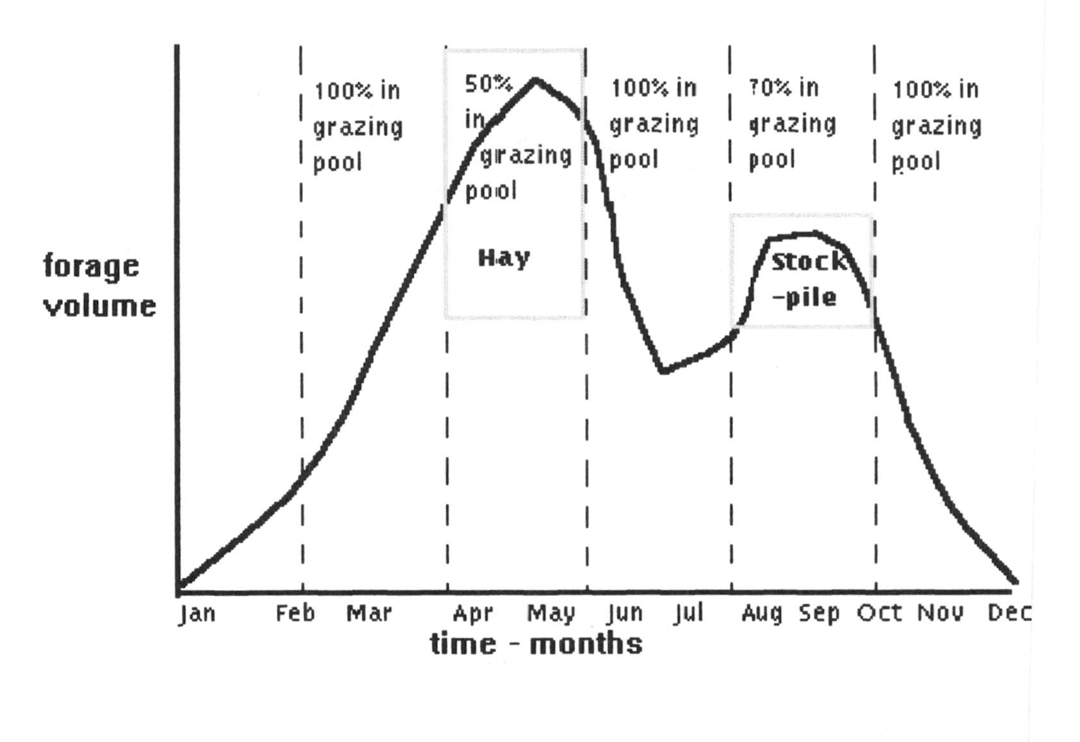

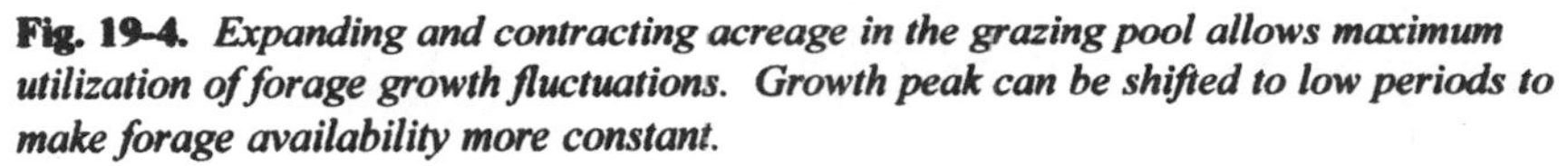
Fig. 19-4. *Expanding and contracting acreage in the grazing pool allows maximum utilization of forage growth fluctuations. Growth peak can be shifted to low periods to make forage availability more constant.*

mineralized due to dryness) it does not take as many pounds of forage to get the same nutritional kick as it did in the washy, wet spring.

This increased nutritional strength in the grass of course works in our favor because growth is declining. Adding in the hay ground not only increases the acreage in the grazing pool but also increases the number of paddocks. Instead of having 20 paddocks, like we did the first rotation in the spring (remember we wanted to cover the whole farm in 14-21 days?) and perhaps 25 in the second rotation, now we have 50. By increasing the number of paddocks, we've increased the rest period between grazing. As grass growth slows, we need more rest because the recuperation period increases as growth slows.

In a serious drought situation, we may need 100 days of rest. If we run out of paddocks, we feed hay rather than eat the grass into the ground. This technique routinely nets more benefit than a hard shock to the grass. The worst thing to do in a drought is to denude the soil through overgrazing. Taking off the ground cover opens it up to further drying, shortens roots even more, and reduces biological activity. It also retards regrowth once rains come.

Several times, during extreme droughts in late summer and early fall, we have fed hay for a few weeks or even a month to shepherd the grass. As a result, when others in the area ran out of grass and began feeding hay for the winter in October, we were able to go back to stockpiled forages and not feed hay until Christmas. In other words, by protecting the grass during a drought, it came back with such vigor once rain finally did come that feeding hay for four weeks picked up eight weeks worth of early winter grazing. The net gain was four weeks' grazing.

If we had gone ahead and eaten the grass into the ground, we would have fed hay four weeks longer during the winter. This is one of the hardest principles to get across because it runs counter to conventional wisdom. I've had cattle farmers tell me: "Well, those cows are just going to have to survive, because I'm sure not going to give them any hay in July. Who ever heard of such a thing?" Taking a little pressure off the grass, though, would net them tremendous increases in production down the road.

Some of the best land improvement can be done during a period like this. A drought provides opportunities to graze around a pond because the ground is hard. They will not damage the edges or backside of the dam like they might during normal times. It's a good time to graze a swamp and open it up for more species diversity. Even running them through a woodlot to prune low branches and do some minimal light aeration can be beneficial during this time.

Brushy areas are ideal sites for hay feeding. Any area where grass is having a hard time growing, whether it is because of exposed soil or weeds and brambles, can be turned around completely with some hay feeding during this time. It's amazing to throw some hay out on a bramble patch and watch the cows completely stomp out the brambles. The same activity performed in the winter or early spring would pack the soil into brick, destroy any grass that was there and stimulate the brambles. Strategic use of dry times can turn them into assets.

As fall approaches, grass growth usually increases again for a few weeks. We can contract the grazing pool again, using only 70 percent of the acreage, and leaving 30 percent for stockpiling winter feed. By selling all the long yearling calves as beeves during this time, we reduce the cow numbers going into winter when grass growth virtually stops. This again helps to match the intake requirement to the available forage. Since the stockpiled forages deteriorate in quality, we can wean calves and put dry cows on the older stockpiles and get the full benefit out of them.

This grazing flexibility means that we may go into a field one time and graze it in three paddocks (three days) but the same field may be grazed in 15 paddocks later in the year. If that area is used for stockpiling with the dry cows in late December and January, it may be grazed in 50 paddocks. This is why permanent fences must be kept to a minimum. The more permanent fences, the less flexibility.

Certainly permanent paddocks can always be subdivided, but if the paddocks are small enough to accommodate the times you want paddocks small, then when you want them large the cows will be walking around gates like a maze, causing traffic damage around gates. The time of year when this would occur would be in wet times under fast grass growth conditions, which exacerbates the traffic problem.

Hay making becomes much more difficult too when all permanent paddocks are small. The most efficient haymaking is on large acreages.

I've never seen a farm rely on permanent paddocks that didn't overgraze sometimes and undergraze other times. Overgrazing is just as detrimental as undergrazing. Overgrazing weakens the forage due to depletion of energy reserves, pushes succession backwards and does not accumulate as much solar energy because the grass is never allowed to hit the blaze of growth period. Undergrazing mulches down the grass crowns, discouraging tillering and thinning the stand, pushes succession backward through thinning of the sward and fails to optimize solar energy because the plants are beyond the blaze of growth period. Isn't it amazing that the results of overgrazing and undergrazing are practically identical?

Flexible paddock sizing is imperative for what Burt Smith calls "variable density grazing." This is sophisticated terminology for the concept that we need flexibility in stocking, facilitated by paddock size, to fully capitalize on grass growth, nutrient requirements of the stock and time of year.

In the fall, slaughter beeves must receive only ice cream. As a result, we are content to utilize only 50 percent of the available forage. This is what we call "popping" the beeves before processing. For dry cows in the winter, however, we attempt 70 or 80 percent utilization because we are feeding mainly the box the ice cream came in, and only offering a smidgen of ice cream. Stock density — again regulated by paddock size — is the gas pedal and the brake of this flexible utilization principle.

In the spring, with big paddocks, cows can calve and find the corner by themselves for privacy. In the middle of the summer, though, with small paddocks (necessitated by longer rest periods and allowed by stronger grass) the cows would have trouble finding a private spot to have their calves. But the bull has all the cows "close and personal" for efficient breeding. See how it all works together? It is truly wonderful.

While this is certainly not a comprehensive discussion of the topic, I hope it does make the point that grazing management is to the land what a fine sculptor is to a piece of clay. Yes, there are definite principles involved, but in the final analysis only the grazier, with animals, creates a beautiful synergistic whole. May we all aspire to such a noble calling.

Chapter 20

Grass Observations

I've always been amazed at the perennial forage community. It changes according to season, soil fertility, animal impact and probably other factors I don't even know about. Some of these characteristics are downright amazing when viewed as a whole.

One thing I've noticed is that the more fertile the soil, the fewer the seedheads and more numerous the leaves. This phenomenon may be a result of both fertility and animal impaction, rather than just fertility. Since we don't have any areas free of animal impact, I don't know for sure.

But I do know that in any given paddock, a more fertile spot will yield grasses with fewer seedheads and more blades. Psychologists who study populations tell us that big families result from social instability. When food is scarce and life is difficult, families tend to be large so that at least a couple of the offspring will survive and provide for the parents in old age. As a culture's wealth and stability increase, family size adjusts downward.

Could this same principle be at play in the perennial forage community? I suggest maybe it is. When grass is strong and its roots have ample nutrition, it need not make many seedheads. Its future is assured. It puts its energy, instead, into growing more leaves and amassing more solar energy, creating more wealth, so to speak.

The ratio of blades to seedheads goes up as the fertility increases. I do not know how to explain better small grain harvests (seedheads) due to extra fertility. Perhaps one of the differences is that a perennial forage is mainly concerned with volume of vegetative material instead of seeds. Annuals may be genetically programmed to make more seeds.

One practical result of this growth pattern is that late harvesting becomes more possible. The nutritional level, as well as palatability, declines dramatically from the early boot stage (when you can barely see the seedhead down in the whorl, or interior of the plant) through full seedhead stage. Often, protein percentage can drop by 50 percent in this two week period. The protein content of the leaves, however, remains fairly constant. In other words, the lab tests all the biomass on the square foot of ground.

The carbonaceous stems tend to pull down the total protein percentage in the plant. But if there are almost no stems, the nutritional level will remain nearly constant for a long time, perhaps for an additional two or three weeks. That extension of palatability means a longer hay harvest period.

I've seen fields of extremely similar character (soil type, topography, aspect) vary as much as night and day it seems in proportion of seedheads to blades. And these are not just tiny areas. They cover more than an acre. Other observable differences include the number of earthworm castings in the low seedhead areas, and the soil moisture. A droughty area will make more seedheads as a percentage of total plant material, illustrating the importance of organic matter to this whole pattern, for it is the OM that fights drought.

A related aspect is delayed senescence. Senescence, or aging, is similar to seedhead creation but goes much further. And herein lies my argument against planting winter annuals. Under any stress conditions, whether it be drought or cold, the more fertile the soil, the slower the grass ages. I'm not talking about maturation, which is the creation of seedheads, but rather the dormancy and then dying of the grass plant.

The observable part of this change is that the plant goes from green to yellow to brown. It goes, nutritionally, from proteinaceous to carbonaceous. It goes from palatable to repugnant. So this change is extremely important. If we can delay senescence, it means we can delay the time when, under stress, the forage ceases to be a valuable feed source and we are forced to supplement or use stored feed.

I've noticed this most in the winter. All the experts say cool season grasses stockpiled for winter grazing must be used by a certain time. After that time, they deteriorate to the point that livestock won't eat them. Our primary cool season grass is fescue. If it is given an application of manure, and especially poultry manure, in the fall just after being grazed, this grass will stay green right on through the most bitter winter.

Synthetic ammonia applications will not do the same thing. And it won't work if the forage is not grazed or mowed to stimulate regrowth. But fescue not treated with manure will generally turn brown on top by the first of January in our area. Adjacent paddocks that received manure in the fall, however, stay green like spring lush all the way through. Growth volume is the same. Spring regrowth will be about the same too. But winter color is unbelievably different.

In terms of winter annuals, which are planted to keep everything green in front of the livestock during the winter, I believe we could accomplish the same thing by spreading manure ahead of a warm fall rain on short cool season forages.

This technique would eliminate the need to establish a winter annual, which runs anywhere from $50 to $250 per acre, depending on the plant, tillage, fertilizer and herbicides used. In economic terms, this green perennial will beat the annual every time. Livestock will graze it into the ground.

The grazing patterns reflect the plant's nutrition. On areas so treated the animals eat it like cotton candy. On untreated areas in the same paddock, they pick and choose halfheartedly. Even when lounging, they prefer the green area over the brown, so that their manure is applied not to the area that needs it more, but to the area that needs it less. This meshes well with the whole synergistic idea of nutrient cycling.

Another thing I've noticed over the years is that higher fertility plus animal impact encourages better grass species and discourages the poorer ones. In our area, fall panicum, commonly known as redtop, is viewed with disdain as a weed grass. I find that it is by far the preferred species of the cattle during late summer droughts. And it grows in drought along with broomsedge when everything else has gone dormant.

But in areas where we graze and immediately apply manure, regardless of weather, the clovers, orchardgrass and fescue will beat out the redtop and broom sedge compared to an adjacent area not so treated. There, the redtop will dominate and make seed and the better species will not grow until soil moisture increases. Again we come back to the value of organic matter to hold moisture. This whole concept corroborates Voisin's meticulous notations of species succession.

The perennial forage species composition of an entire field can be changed dramatically in one year by manure plus animal impact. We don't need to overseed the good species to make them grow. When their environment is stimulated, they will grow automatically.

We tend to think so many things are impossible without machinery and labor. If management and carefully timed manure or compost applications can lengthen good quality hay harvest by two weeks or more, eliminate the need for winter annuals and bring in the best forage species, that seems to be where we should focus our money and energy. Most capital intensive procedures, because they do not deal with the mismanagement that caused the problem, must be repeated time and time again.

We get goosebumps contemplating what our farms can look like after half a century of capitalizing on natural principles.

Chapter 21

Cow-Days: Measurements and Applications

The cow-day is to the grazier what bushels are to a corn producer and inches are to a carpenter. It is a standard measure and transcends pounds per acre or total digestible nutrients per acre or any other measure.

A cow-day is the amount of feed one cow will eat in one day. Some folks use animal units, which is a more generic term and works for goats, sheep, llamas, what have you. Since I work in the realm of cows, I like cow-days.

The reason this measure is so valuable is because it takes into consideration fluctuating intake. A cow just ready to re-breed and nursing a 200-pound calf will consume far more forage in a day than a dry cow on a warm early March day. In fact, that dry cow on a subzero February day will consume far more than that same cow on that warm day in early March.

A yearling eating grass for compensatory gain will consume more volume in a day than that same calf two months later in a drought. It takes half again as much lush spring grass to get the same nutrients as it does in July. Frosted fescue is often 14 percent sugar, while unfrosted fescue is only 7 percent. That can change in just a few days.

Stocker calves eating hay in the winter will alter their intake by 100 percent depending on quality of the feed. Clearly a cow-day of good hay would be far more than a cow-day of junky, unpalatable hay. Learning to estimate these cow-day values is accomplished through trial and error and experience. My point is that we cannot accurately determine how many animals we can feed based on pounds of grass or TDN (Total Digestible Nutrients) because there are simply too many variables at play.

Quality of the forage, nutritional needs of the animal, weather and perhaps some other things we haven't yet discovered all affect consumption. The cow-day takes into account all these variables. For that reason, the cow-day is the standard. To me it is easier to guesstimate cow-days than it is to take forage samples, measure them for TDN, decide what my nutrient requirements of the stock are, and then adjust the paddock size accordingly. I'd be afraid the cows would starve to death before I got it all figured out.

I keep a running herd cow-equivalent in my head. In other words, if we have 25 mature cows, 25 month-old baby calves each worth 1/5 of a cow, and 25 yearling calves each worth half a cow, we have 25 + 5 + 13 (12.5 to nearest one), which equals 43 cow equivalents. In the winter if we have 40 mature cows and 40 stockers, each worth half a cow, we have 60 cow equivalents. If we buy thirty 500-pound stocker calves, we have 15 cow equivalents. But as they grow and average 800 pounds, then we have about 24 cow equivalents.

Certainly such figures are somewhat subjective, but they are easy to work with on a day-to-day basis. It doesn't require a lot of pencil-pushing and calculation to run these numbers. A herd is never static, but changes cow equivalents through weight gain or sales additions and subtractions.

The other part of this figuring is to tie it to the grass. If we have 50 cow equivalents on one acre for one day, we harvested 50 cow-days of grass during that stay. The formula is:

$$\frac{\#\text{ COWS} \quad \text{X} \quad \#\text{ DAYS}}{\#\text{ ACRES}} = \text{COW-DAYS PER ACRE}$$

Make sense? If our 50 cow equivalents spend 10 days on 5 acres, we have 500 cow-days divided by 5 acres: we harvested 100 cow-days worth of forage per acre.

We use the cow-day measurement many ways. The fundamental use is for determining the size of a paddock. Let's say we have 50 cow equivalents and we want to harvest 20 cow-days per acre in a one-day stay. We would need 50 cow-days divided by 20, or 2-2/5 acres. Since an acre is 4,840 square yards (I just figure 5,000 to keep it easy) we would need 12,000 square yards (2 times 5,000 plus 2/5 of 5,000). Follow me?

Once you get a feel for how wide your fields are, allotting the acreage becomes simple math. If your field is 200 yards wide and you need to offer 12,000 square yards, all you need to do is step out 60 yards and run the new break fence. Setting up paddocks by eyeballing the field is as inappropriate as using permanent fence for all subdivisions.

With nothing but permanent paddocks, one day you will not offer quite enough and the next day you may offer too much. One day the cows are short, and therefore dissatisfied and lower in performance, while the next day they waste grass — both scenarios affect our profit and, in a more theoretical sense, your conversion of solar dollars into usable product.

On some extensive ranch acreages, temporary fencing is not possible, and I hope folks in brittle environments will forgive some of this clearly temperate region formulating. But I never cease to be amazed by the number of graziers who do not use cow-days. Again, because dry-cow-days are different from lactating-cow-days, we can make paddock adjustments efficiently and accurately using this method. I can't imagine trying to lay out paddocks for efficient 24-hour moves without a standard measurement. Furthermore, the figuring needs to be simple enough to be done while I'm picking up yesterday's fence stakes.

The applicability of this formula is universal. Large operators would gain a sense of acres, while small operators develop a sense for square yards. For example, if you had 5 cows and you had

50 cow-day grass, a one-day stay would be 1/10 of an acre, or roughly 500 square yards. See how versatile and easy this is?

Literally from one day to the next, cow-days can shift. For example, we can move from an exceptionally lush, fertile field at 60 cow-days to an adjacent hillside low-fertility field of 20 cow-days. How do you know how much to allot the animals if you do not have a feel for how many cow equivalents you're feeding and how many cow-days are in the next field? Standing there and squinting at it for awhile will not do the job.

Monitoring becomes easy using the cow-day measure. If you notice that you missed it today and want to correct tomorrow, you can quickly down-size the next paddock the appropriate amount. Did you oversize by 10 percent, 5 percent, or perhaps 20 percent? Deleting that many more square yards (or acres) is easy. Perhaps you sold some animals and dropped by 20 cow equivalents. What are you going to give them now?

Because we keep a grazing schedule, we can look back at the end of the season and add up the number of days in different pastures, multiply it by the cow equivalents per stay, and divide by the acres to get a total cow-days per acre for different fields. This is a fabulous tool to monitor productivity increases or decreases. If plotting annual cow-days reveals an escalating line, we know we're heading in the right direction. If it's going the other way, we know we're going backwards. Sometimes it can go up on one field and down on another. This way we can catch small problems before they become big ones.

These records also help us adjust our goals. For example, if we have a field giving 300 cow-days per acre per year and an adjacent field of similar characteristics giving only 225, we can confidently set 300 as a goal for that field. It gives us a reachable goal, but with a way to specifically measure progress. I would hate to think that the only way I could monitor the farm's progress was by eyeballing it.

This is also an excellent planning tool. Since most graziers keep a grazing schedule, updated every couple of weeks, the cow-day history of a field provides precedent for cow-day budgeting. Cow-days are the grazier's dollars, and if a field has a record of yielding 200 cow-days during its second seasonal grazing, chances are it will yield a similar amount the next year. If we've added stock, changed stock type, or reduced stock, we can plug the current cow equivalents into the 200 cow-days and come up with the number of days' worth of grazing in that field. Dividing the acreage into those days gives us the exact size to make the paddocks.

Perhaps one of the most important uses for the cow-day is in determining number of grazing days left in the winter or in a drought. When we enter a drought, I walk the fields and note how many days of grazing we have left. For example, judging by cow-days (remember in a drought cow-days per volume goes up because the grass is stronger) I'll make a note that field A, which is 5 acres, has 45 cow-day grass so that's 225 cow-days there; field B, which is 8 acres has 30 cow-days grass, so that's 240, for a total of 465 cow-days, and so on until I come up with a grand total.

I know we have X number of cow-equivalents, so I divide that into the total available cow-days and have a total number of days we can graze without any more grass growth. I'll often come in and tell Teresa: "Well, we can go 12 more days without rain." The same system works well in the winter on stockpiled forage. We don't have to worry if the cows are getting enough to eat. We can monitor utilization daily and measure the amount of acreage of that same cow-day type feed out ahead of them and know that we have so many days of grazing left.

The value of this monitoring and budgeting cannot be overestimated. It keeps us from worrying about what is going on. We are not caught by surprise. Remember that cattle slide backwards long before we can see it by observation. If we are just putting those cattle out there and hoping they are finding enough to eat, by the time we realize they are not we've lost many pounds of gain. But knowing what our budget is and systematically offering them that fresh salad bar every day allows us to keep productivity up and not bite off our fingernails even in rough circumstances. For emotional stability alone it is worth every mental computation.

By walking the fields every few days, we keep a mental note of how the grass is growing, which paddocks need to be grazed, which ones we overgrazed or undergrazed; and by constantly recording the cow-days, we can fine tune the stock density to capture every possible unit of solar energy.

When figuring carrying capacity for your farm or another, the cow-day formula is an invaluable planning tool. Once you develop the expertise to walk onto a farm, look at the sward, and give a reasonable estimate of cow-day productivity, you can get a fast and reasonable ballpark figure for carrying capacity. For example, if you walk on a 100-acre farm that has been under continuous grazing with an annual average of 80 cow-days per acre, you can be fairly confident that under management-intensive grazing it can go to 160 cow-days the first year and gradually get to 250 or better within five years.

You can use this information to decide how many cow-calf pairs it could handle, or how many stocker calves, or how many dairy bull beeves, or dairy heifers — the list could go on and on. The point is that this knowledge is the beginning point for running some figures for potential production and comparing one enterprise with another if you are going to lease it or buy it. Again, the centerpiece is the cow-day. If we do not have a basis for determining what the carrying capacity is for a piece of ground, certainly we do not have a basis for deciding how much we can afford to pay or even if it will cash flow.

This is where the experienced grazier can out-negotiate a conventional cattle operator. Once a grazier knows what kind of cow-days can be obtained from a certain type of forage on a certain type of ground, he can make what might appear to others to be an outrageous decision. But the experienced grazier, armed with cow-day knowledge, is armed with the best weapon in the world: good information. The foundation of graziering information is to utilize the cow-day measuring stick.

Another application of the cow-day is in nutrient analysis.

For example, we know that a cow is dropping 50 pounds of material out her back end every day. That means if we have a field that gave us 300 cow-days per acre this year, we applied 15,000 pounds of pasture droppings. Certainly not every ounce of nutrients was metabolized by the soil and assimilated by plants, but it does give a rough idea of what has been applied. Since every cow-day produces a third of a pound of nitrogen, those 300 cow-days returned nearly 100 pounds of N to the land.

This type of analysis is helpful for farmers contemplating fertilizer purchases and truly points up how much money is spent in the wrong places. It is typical, for example, to total up the cow-days on a farm, then multiply it by the 50-pound-per-day rule and come up with an average of five tons of manure per acre per year. Clearly something is wrong if this farmer spends $10,000 a year on fertilizer. And yet this happens routinely.

The weak link is not fertilizer; the weak link is utilizing the resources already being produced. The manure could be applied in the wrong places, could be translocated, or could be leaching and/or vaporizing due to inappropriate handling. But converting total cow-days is a way to get a handle on fertilizer being returned to the pasture and making better soil amendment decisions.

Finally, the cow-day is an excellent financial planning tool. Especially in a brood cow herd, the income from calves must be spread over the cows. To say that a calf "made me X amount of money" without figuring the cows into the equation gives an inaccurate view of the herd finances.

I like to divide total cattle income by the total number of cow-days per year. This can be tricky because the cow-days fluctuate throughout the year. But take the months at a certain cow-equivalent and get a total for that month, then the next and the next and finally add them all up. It's a good guesstimate. The idea is to get a total income per cow-day.

Now you can do the same thing on expenses and eventually get to a total gross margin per cow-day. This is one of the best planning tools I know, because it can enter into practically every decision we make.

For example, if our gross margin is 10 cents per cow-day, we would not want to do anything that would reduce that amount. Preferably, we would do only things that would increase it. For example, let's assume we have a shortage of feed and that we can purchase grain and increase our cow-days but the gross margin is only 9 cents. Instead, we have an option to buy hay but our gross margin is only 8 cents. But we could sell some cattle and end up with a gross margin of 12 cents. Obviously, that would be the better option.

If buying fertilizer will increase our cow-days by 10,000, say, at a cost of $2,000, but building fence and tightening up the grazing will increase our cow-days by 5,000 at a cost of $500, one is adding cow-days costing 20 cents and the other is adding cow-days costing a dime. If our cow-day cost average is 20 cents, clearly the 10-cent-per-cow-day expenditure is a good investment.

This kind of analysis works great for deciding how long to keep cattle, whether or not to background calves with silage, and figuring boarding rates. Let's say my gross margin per cow-day is 30 cents and I have extra feed. Since I don't want to sell it because that is like selling fertility, I want to put that extra forage through cattle. Do I board or buy or run on gain (grow someone else's calves for a share of the pounds gained)? If I can buy and make a gross margin of 30 cents per cow-day I would probably want to charge 30 cents per cow-day to board my neighbor's cows. On the other hand, if the neighbor comes and offers me 50 cents a cow-day to board his animals, I'll jump at the chance.

Once you have a cow-day margin figured, it's an emotional shot of adrenaline each day to know that "those cows made us $35 today." When paydays are spaced so far apart, prorating the cow-day margin helps us enjoy daily chores, knowing what our paycheck was for that day. This may sound esoteric, but the emotional element of sticking with our work when no weekly reward comes is very real. Breaking down the big picture to 365 little pictures helps satisfy this emotional need.

For many reasons the cow-day measurement is ***the*** standard for the grazier. It is flexible, versatile and understandable. The more you use it, the more you will come to depend on it for information, planning and decision-making. Perhaps if enough of us begin producing salad bar beef it will become a household term like inches and bushels.

Chapter 22

Moving Cattle: Logistics and Economics

A lot of people say to me: "The way you farm takes too much time."

It's important to remember the paradigm under which most livestock producers operate. When I say I move cattle every day to a new pasture, most cattle folks have a mental image of what I've just said that confirms their judgment that I do "too much work." For most cattle folks, the phrase "move cattle" is equivalent to threatening them with a migraine headache. Around here, when folks "move cattle" it means they get a couple or three pickup trucks, a couple or three four-wheelers, a couple of dogs, several cans or pouches of Skoal or Red Man, and spend most of a day beating the brush, cussing, fuming, yelling, and generally having a frustrating day. At the end of it, they probably still left a cow back in the "bresh" (brush).

That paradigm defines what they think when I say I "move cows" every day. Such a notion is like telling them I enjoy amputating my fingers. They figure I must be a glutton for punishment or a masochist. To say that "it takes too much work" is just their polite, courteous way of saying: "You're crazy."

Of course, the time and work argument could not be easier to refute. My response is: "Well, I don't have lights on my tractor." The fact is that I guarantee the salad bar beef system is much less time consuming and returns far more money per hour than conventional livestock farming.

Photo 22-1. *"Come on, cowies!" Cows learn quickly to heed the daily call to a fresh salad bar.*

When I say "move cows" I mean to walk out to the paddock, open the cross fence, step back and call the cows. In five minutes they've all zipped through and started grazing in the next paddock. I close the cross fence, take up the prior backfence, roll it out for tomorrow's new break fence, and walk to the house. We literally spend no more than 30 minutes a day "moving cows."

After the cows are trained to the electric fence, it's amazing what you can do with them just stringing up what I call a "bluff" fence. If you need to make a turn, go around a hill, or go across a field to a farther one, you can just stretch up a bluff fence and it will be as good as a permanent hot wire. Many times this is simpler than making complex interconnecting lanes everywhere. Main lanes certainly have their place, but sometimes you need a lane where you don't have one or you need to make a turn and you don't have another person to stand in the way and help. A bluff fence works great and just takes a few minutes to put up.

I like to move the cattle in the late afternoon because they will graze longer into the cool evening before lying down. The cows

Photo 22-2. *We open the cross-fence to let the cows enter fresh salad bar. Note the graze line between the old and new paddocks. Electric fence works.*

generally begin grazing immediately upon entering a paddock and graze until they either get uncomfortable from midday heat or until they fill up. If I move them in the morning, they generally want to quit grazing when the sun gets warm. Furthermore, the dew gets burned off and they drink much more water.

They walk more because they graze during the hot day when they naturally either pace around the paddock more or lie down under shade. By moving in the late afternoon, around 4 or 5 p.m., they graze aggressively up until almost dark and then lie down, completely satisfied. They graze during the night on heavily dewed grass, reducing their water consumption. Their heavy rumination occurs during the cool night instead of the hot day.

In the winter, on stockpiled forages, the same time seems to work best. Cattle do not like frosted grass. If I move them early in the morning on frosted grass, they trample a lot of it because it is not as palatable. But if I move them in the evening, they graze well up into the night and by morning, when the frost is on, most of the grass has been grazed off and they have their bellies full. As a result, they

Photo 22-3. *Here they come, looking for the most palatable forages first. "Moving cows" is child's play.*

will often just continue lounging until late in the morning when the frost melts. The same principle holds for early spring grazing.

In the spring, the added benefit of not breaking over frosted grass comes into play. Bending over and breaking frosted spring grass is hard on it. It does not grow back as fast and discolors. Again, the late afternoon move reduces this damage.

Moving across the pastures requires a minimum of three cross fences. Two provide the boundary for today's paddock, and one defines the new break fence where they will go tomorrow. I never like to leave the cows without the new cross fence in place. Once in awhile a deer will run through the wire or a kink will weaken the wire and it will break. In either case, the cows go into the next paddock. If there is no break fence to stop them, they can go all the way to the other end of the pasture and do a tremendous amount of trampling damage.

If the back fence happens to go down, the cows do not go back because there is nothing there to graze. A safety on that end is unnecessary. A break occurs only a couple of times a year, but when it does, you want to be ready for it.

Photo 22-4. *Keep the wire tight and let the cows come around the end. This daily look-over is invaluable for monitoring the stock.*

We always keep the spark out of wires we are not using. That concentrates the spark in the wires where the cattle are, making a hotter spark, as well as making it easier to trace down a short. It also reduces the likelihood of a short because very little of the wire is hot.

If you need to go back to a water trough, a lane 8 feet or less in width will keep the cattle from lounging there. Cattle do not like to be too close to an electric fence while they are lounging. If you narrow the access lane to the water trough, they will walk down to drink but immediately turn around and come back. This is the key to eliminating loitering around the water tank. I'm sure there are some cows out there to prove me wrong, but they are the exception, not the rule.

If you bring in new animals, certainly a training period in a corral with electric fence is good. But if you back the trailer over an electric fence and dump them in with your existing herd of well-trained animals, they will train by default. First of all, they will want to fight with the old group rather than run around exploring (and consequently break through the electric fence), and this keeps them from running a long distance — running through the electric fence.

Photo 22-5. *Stay out of the way and let the animals come. They look forward to a new salad bar just as eagerly as people do. The move is soon complete, without yelling, chasing, or any of the typical behavior associated with "moving cattle."*

Since the trained animals naturally want to stay away from the fence, the fighting between the old group and newcomers generally occurs away from the fence too.

Secondly, the newcomers are no doubt hungry. If you put them in a lush paddock, they will be more interested in eating than in running. By the time they are interested in exploring beyond the fence, they've settled down and they'll hit the fence gently. The key to training is to get them to hit the fence gently — preferably nose it, inquisitively. That provides the most memorable experience and the quickest training.

If the paddocks are thin, like in a strip-grazing period, it's easier to move the cows by rolling up the break fence than by opening it up with the gate handle and asking them to file down around you. What happens is the first ones come into the next paddock and walk past you toward the end of the long narrow paddock. The cows at the end of the line in the previous paddock see the cows in the new paddock pass them going the other way and do not continue filing toward the open end but turn around and follow the inside cows out toward the end of the paddock.

In such situations, it is much easier to roll up the break fence and let the herd filter in behind you. If you want to put it back, you can just roll it back out and refasten it. The extra time winding up is well worth it.

Anyone who implements controlled grazing successfully knows that moving the animals is the easiest, most enjoyable part of the operation. Just a call or whistle and the well-trained animals come running for their new salad bar. One time the cows rubbed down a gate and got out onto the road when I was gone (that's always when these kinds of things happen). By the time Teresa saw them, they were heading down the road past the neighbor's farm. She went to the gate and called them. They all stopped, turned around, and came right back into the field. She closed the gate and that was that. I think that's when we replaced that gate.

The big question is: "Does it pay?" In our county, the average cow-days per acre of pasture production is 70. That means that our 300 cow-day performance is 230 cow-days above average. If we prorate that additional productivity to a per-move average, it comes out to about 38 additional cow-days per move. Our cow-day gross margin is 80 cents (cost is \$.30 and income is \$1.10).

The net income value of 80 cents per cow-day, multiplied by the additional number of cow-days accrued from the pasture because of one average daily move, is \$30.60.

That is an hourly return rate of \$60. How is that for return to labor? My response to the economics questions is this: "I couldn't afford to farm if I didn't move them every day." Even if my figures are high, it's still a tremendous return. If we had more acreage and a larger herd, the hourly return rate would be much larger. It doesn't take much longer to move 200 cows than 100 cows: just a few extra minutes to let more of them go through the gate.

And I did not add the fact that the dollars saved are worth \$1.30 because what you save you don't have to earn and pay taxes on. In other words, saved dollars are worth the tax difference more than income dollars. Other folks plant corn silage and buy fertilizer and build silos to compensate for their poor pasture production. In other words, they get 800 or 1,000 cow-days per acre off cultivated crops in order to compensate for poor pasture productivity.

I believe this illustration gets to the heart of the "pleasant life in the country" idea that so many folks dream about. It's not how many hours we put in, or how hard we work. . . necessarily. It is how we incorporate the different components of the farm to capitalize on nature's laws, orchestrating the different pieces in symbiosis. This way the plants and animals do much of the work, and we can lie down at night thinking about all those earthworms out there busily aerating the soil and making trace minerals. Isn't it exciting?

Don't be bullied by folks who think animal movement doesn't pay. Just go out there and do it, and enjoy it so much the naysayers will shake their heads in bewilderment. Go make some money: move those animals.

Chapter 23

The Mob

How many cows does it take to make a mob? Controlled grazing, at its most fundamental level, seeks to mimic the animal behavior, forage defoliation and soil disturbance of huge wild grazing herds.

These vast herds are numerically far more than even the largest American cow herd. It is a mistake to think that mob effect is simply a matter of pounds of liveweight per acre. At some point, a group of individuals becomes a mob. In addition to space intensity, a working "mob" needs numerical magnitude.

To illustrate, think in terms of a dozen English fans crowded in a room watching a soccer game. We all know those folks take soccer seriously. Now imagine 10,000 fans, packed just as densely (i.e. same square footage per person) in a stadium watching the same game. The latter destroys retaining fences and goal posts. The latter is a mob. The former is a group — rowdy perhaps — but nonetheless limited in destructive potential.

With the onset of winter, and far too much feed for our own herd of 50 plus animals to consume one winter. I contacted a neighbor who was short of forage about wintering some of his dry cows. He most happily agreed with the plan and I soon found 79 dry cows on my hands.

Because they came in two groups spaced roughly two weeks

apart, I enjoyed watching a herd of 39 for awhile before getting the other 40. The first group consumed stockpiled fescue at a little higher cow-day figure than I had planned, but not enough to astound me.

When the second group came, I amalgamated them into one large group and ran them in the same field and adjacent to where I had been strip grazing my own small group of cows. I was running my cows at about 30,000 pounds of liveweight per acre, moving them at three day intervals. I was getting roughly 50 cow-days per acre on fescue that had been mowed twice for hay during the summer.

I moved the large herd into adjacent paddocks at about 30,000 pounds of liveweight per acre and suddenly the cow-days jumped to 80. Same field, same forage, same pounds per acre, but a whopping 35-plus percent increase in cow-days. The only difference was that one was a group, the other a mob. Same density, different magnitude.

Several differences were quite apparent. The group grazed in a more scattered, individualistic pattern. There may be a cow at one end of the paddock and another at the other end. They seemed to like their territory and tended to put large spaces between themselves. The mob grazed altogether. The followers grazed right on the heels of the leaders. They didn't look up to see where their neighbors were. With nose to ground, they just hogged it in. They grazed all the way across the paddock, slowly, then turned and came back.

When I walked out to take mineral or check them, or to move them, the mob was far more easily excited. I don't believe all of this difference was because my cows were familiar with me and the others were not. When it came time to move, the mob cows kicked up their heels, pranced sideways and butted heads. The group, however, just watched to see where I opened the fence and walked to it.

The mob exhibited much more patience. The group, clearly composed of individuals, would begin pacing the paddock. Then one cow would bellow her complaint: "It's time to move." The mob, on the other hand, would wait patiently, not one making a sound, even when their forage was eaten far more thoroughly than the group's. I don't know if they were not as hungry, although I would think they were. Perhaps the peer pressure was too great for one to voice her disgust.

Clearly, the mob utilized a greater percentage of the available forage than the group. Trampling wastage was virtually nonexistent. Hoof chipping on the soil surface seemed double that of the group, which tended to move with more visual caution than the mob.

Based on the residue from the group paddocks, it was apparent that the entire 35 percent increase in the mob's efficiency could not be attributed solely to amount of available forage harvested. Could it be that individuals in a mob require less feed to be satisfied? Or stated another way, perhaps a mob requires less feed for identical performance.

What does all this mean to the grazier? A couple of ideas come to mind. The most obvious is that all things being equal, the larger the herd, the more efficient it is. Communal grazing could offer some real advantages. If three graziers running 50 cows apiece could put together a mob of 150, each could possibly increase his herd by 17 or 18 cows and graze the three farms intensively as one whole.

Another interesting possibility is that creep grazing calves ahead of the cows, or running groups in a leader-follower arrangement may not be as efficient as some think. In nature, the calves, yearlings and mature animals all run as a herd. Perhaps amalgamating even these diverse subgroups would increase the efficiency of the whole more than the efficiency of one group (i.e. average daily gain of yearlings or calves). In other words, what is gained by giving the leaders "first dibs" may be lost by having two smaller herds. Or put another way, the average daily gain increase on the calves may not pay for the overall herd efficiency decrease caused by having a smaller mob.

Most of the research justifying leader-follower programs is based on performance of the leaders. Whether it is milk production in dairy cows or average daily gain in stockers, the benchmark is always performance of the more nutrition-demanding group. But this monitoring does not take into account the loss in mob character

— numerical magnitude — that occurs when the herd is split into two groups. Perhaps the loss in total production, or total cow-days, negates the production gain for a part of the herd. And all the while the work load doubles because of the two groups.

At what point does a group become a mob? Based on my observations, I'd say it is somewhere around 50 cow-equivalents. Below that, the ability to achieve mobbing instincts are drastically reduced, regardless of how confined the grazing. And the mobbing benefits increase as the numbers increase.

Chapter 24

Monitoring: Watch the Cows

Since much of this graziering is art rather than science, the questions inevitably arise: "How do I know the cows are getting enough to eat?" "How do I know I've overgrazed?" "How do I know when the grass has reached energy equilibrium?" These and many more occupy hours of discussions in graziering groups. And, honestly, no one knows all the answers.

But there are some rules of thumb, and the foundation of them all is this: ***watch the cows***.

Cows will only graze a maximum of about eight hours per day. Since cows only gain weight and produce milk when they are ruminating, lounge time must be maximized and grazing time must be minimized. Grazing is hard work. Grab a handful of grass and yank it until it breaks off in your hand. It'll make you appreciate how much energy that cow expends when she's grazing.

If you see the cows grazing five or six hours a day, something is probably wrong; if you see them grazing eight hours a day, something is definitely wrong. Ideally, at least in non-brittle environments, they should graze only two or three hours a day. Under perfect conditions, a cow can fill up in less than an hour. Young growing stock on highly palatable forage will graze longer than cows on less tender material.

If the paddock is over-rested, you can see quite a bit of trampled, long-bladed grass adjacent to well-grazed areas. If there is no residual at the end of 24 hours and the cows are still wandering the paddock, you can be sure they did not get enough to eat.

If the cows pace back and forth along the break fence, they are clearly dissatisfied. In this situation, if the paddock is heavily grazed, then they probably did not have enough available forage — increase the next paddock. If the paddock is ***not*** heavily grazed, it is not palatable.

Poor palatability can be caused by overmature grass — perhaps tillers that did not get grazed the last time and now are old. It can be a species problem — solid fescue in the middle of summer is commonly unappealing. It can be a problem with taste, as in bitter forages caused by too much acidulated fertilizer applied — including synthetic nitrogen. Remember that organic matter sweetens forages and chemical fertilizers embitter them. I've seen naturalists analyze soil samples by taste. Yes, it struck me as odd too, but the tasting sense is acute and can pick up nuances that a laboratory full of equipment cannot.

Pacing and bellowing are often symptomatic of inconsistent paddock shifts. Cattle are creatures of habit. They do not like being moved one day at 7 a.m., the next day at 2 p.m., the next day at 5 p.m., and one time being on a paddock 12 hours, the next time 24 hours and the next time 36 hours. They like to be moved about the same time every day (don't you like your meals about the same time every day?) and they like to stay for the same duration in each paddock.

The cows need to learn that they can trust the grazier. If they know that they will get moved today at 5 p.m., they will be less likely to complain when they are a little hungry at 2 p.m. Rereading that statement makes me laugh to imagine what a conventional cattleman would think upon hearing such lunacy. But it is true. The trust between man and beast is just as real with cattle as it is with horses. Nobody pokes fun at the stories of man and horse relationships. Why is it so ludicrous to think cows cannot respond similarly under like care and attention? Of course, we graziers know they can — and do.

Watching cows' behavior is only half the monitoring equation. The other half is watching their manure. If it squirts, they are stressed. This can mean the grass is too immature — back to the candybar diet. They need roughage to keep their rumen balanced. Runny manure is a quick indicator that the grass is overfertilized, under-rested, or too weak. Weak grass normally occurs in the spring, but it can also occur in the summer after a period of heavy rain. Barring these aberrations, it is a good indicator that the grass has not rested long enough.

Manure patties should look like a pumpkin pie. They should stand up a couple of inches in a nice circle, sunken in the middle and built up around the edges. The droppings should splatter a little bit, but not much.

If the manure comes out in ringlets, like donuts, then the diet is too high in roughage. This often occurs when cows are being fed hay, or it can occur if the forage is over-rested and has gotten too mature. Generally, though, in this situation the cows will exhibit obvious repugnancy to the forage and you will have trouble getting them to eat it. On hay, it doesn't seem to matter as much.

If you call the cows and they are reluctant to get up and come, you probably gave them too much grass. If the cows mill around all afternoon waiting for you to move them, you should allot them a little more.

Building relationships with the cows is one of the most enjoyable aspects of graziering. When the cattle are eating in the hay shed during the winter, I'll often just lie down on the hay and watch them in the feeder gate. In the summer, I often go out a half-hour before dark and walk through them. It is cool, they are chock-full of grass and completely comfortable, and I can pet many of them. I love to sit down or lie down in the grass and let them come up and lick me with those sandpaper tongues.

I can talk to them gently at such times and it almost seems that they enjoy it as much as I do. This may sound like the confessions of a psychotic, but I think there are millions of people in the world who yearn for an opportunity to do this even once. And if I neglect enjoying these opportunities, I have failed to be a husbandman in that Biblical sense of the word. The person who has opportunities and neglects to utilize them is worse off than the person who doesn't have the opportunities at all.

Because this type of farming does not build on tractor hours, diesel fuel and chemical application, it leaves time to develop observational skills and relationships unheard of in conventional settings. I highly recommend it.

Maintaining the Salad Bar

Chapter 25

Minerals

For going on two decades now, the only mineral we've used for our livestock is kelp, which is dehydrated seaweed. Kelp has certain properties that makes it an ideal mineral supplement.

The primary advantage is that the minerals are in natural balance. The organic or natural approach points toward relationships between components as being more important than the volume of any single component. In other words, calcium is important, but not as important as its volume in proportion to other elements. The fact that we have X number of pounds per acre of calcium tells us nothing; what is important is how that value stacks up against the volume of other components, especially magnesium, which complexes calcium in the soil.

Balance is the key to stability and productivity, and it is the hardest thing to maintain on any farm — or in anyone's personal life. As important as discipline is, it still should not override flexibility and good humor. All the emotions are good; imbalances cause us trouble. This is why I am always dubious when laboratory academicians prescribe carefully formulated concoctions. First, they seldom ask the animals and second, these are arbitrary formulations based on what man perceives the need may be. And the needs vary greatly from farm to farm, season to season.

To encourage mineral assimilation, the components are either acidulated or synthetically formulated so that the animal must metabolize the material, regardless of need. Lest you think that all nitrogen is the same, or all iodine is the same, be assured that it is not. For example, synthetic urea is an atomic weight lower than the urea that comes out the back end of a cow, according to Charles Walters' *An ACRES USA Primer*. How important is that neutron? I don't know, but I have a sneaking suspicion that it's important. For example, synthetic vitamin paks are far more unstable and subject to potency loss than are identical organic vitamins (i.e. vitamin A synthetically formulated as opposed to alfalfa vitamin A).

Perhaps one of the best illustrations of this point came out of pinkeye research, where iodine was isolated as the deficiency that allowed cattle to get pinkeye. Researchers began giving injections of iodine, and killed as many cattle as they cured. These trace mineral requirements change from day to day, farm to farm, and they could never regulate it closely enough to keep from killing animals. As a result, some states actually banned iodine-fortified mineral supplements, and the industry was afraid of them.

Enter kelp, which has a very high amount of iodine: 1,987 parts per million. In organic form, this type of iodine can be metabolized if needed, but bypassed and excreted if not needed. We haven't treated a case of pinkeye since we began using kelp nearly two decades ago. Neighbors had blind animals who rubbed noses with ours across the fence, but we didn't have the problem. Most of our neighbors are now using kelp.

We prefer Icelandic, geothermally dried *Ascophyllum nodosum*, trademarked Thorvin kelp. Other kelps, especially the ones dried with natural gas, do not contain the mineral levels of this type. Other kelps — many different varieties exist — have varying percentages of minerals. Generally, kelp has 53 trace minerals, vitamins and plant growth regulators called cytokinins (could these compete favorably with growth-enhancing drugs, but without the negative side effects and public perception? Maybe).

The important thing to remember, though, is that kelp assimilates ocean brine directly into its tissue, so that generally the nutrient balance of the ocean brine is mirrored in the dried kelp. Ocean brine mineral percentages are almost exactly the proportions found in healthy human blood. See how the circle closes? Indeed, everything does relate to everything. Generally the cows consume about 1 oz. of kelp per 1,000 pounds of liveweight per day. It comes in a 55 pound bag and greatly enhances the health and performance of the animal.

Several other farmers in our area cooperate in getting a 20-ton container load at a time, saving more than 30 percent off the retail price.

We mix it 50-50 by weight with coarse mixing salt (iodized salt is more expensive). This way we put our money in kelp instead of salt. We keep it in front of the cattle year-round free-choice. Sometimes they eat it voraciously, and other times they don't touch it for two weeks. This insures that they can have it when they need it, but we aren't throwing money down the drain by forcing them to eat it when they don't need it, like a Total Mixed Ration approach.

To avoid reducing kelp's efficacy, certain practices should be avoided. Like making good compost, we need to consider how we handle the manure, how we mix the right amounts of carbon, nitrogen, air, water and microbes. Just because you pile some material together does not mean you will have a successful composting/nutrient cycling program. Just because you put kelp out for the cattle does not mean all your troubles will be over. We have found definite practices that make kelp programs fail.

First is the feeding of any ammoniated feed, or using urea liquid protein supplements and lick tanks. Injecting synthetic proteins and nitrogens into any livestock enterprise, especially when they are fed directly to the livestock, is a dangerous practice.

We've only had one acquaintance over the years who could not get his animals to eat kelp, even after mixing 50 percent dried molasses with it. One winter we boarded two 40-cow groups for him on our extra stockpiled fescue, and asked if we could run an experiment. He was curious, too, to see if we could get his cows to eat kelp.

He was using a high-falootin' laboratory mix from one of those big independent mineral conglomerates. By mixing in a hefty percentage of dried molasses, the mineral smelled sweet and good. When the first group of 40 cows came in, we put our box with the kelp/salt mix right against their box of mineral. By the third day, those cows went bananas. We have yet to bring in animals from another farm or stockyard and not have them go through a three to seven-day binge on kelp within a couple of days after arriving. Whether they come from good farms or poor farms, whether they were on mineral programs or not, this binge pattern never changes. The same thing happens with sheep.

Anyway, the cows binged for about three days and then immediately dropped back to the normal one ounce per head per day (during the bingeing period, they can eat more than eight ounces per head per day). They refused to even touch their old sweetened mineral mix.

Then the second group of 40 cows came and duplicated the procedure. A couple of weeks later, we got a deep snow and had to shift to hay. Our deal was that this fellow would feed hay if it was necessary to do so. He began bringing down his round bales, which he puts under plastic and ammoniates. The hay smells like bleach.

Immediately the cows stopped eating kelp. It was as dramatic as if we turned off a faucet. After a week, I offered to feed hay for a few days because the snow was staying; the temperatures were way below zero, hardening the snow. As soon as I began feeding our hay, the cows binged again on the kelp.

We fluctuated through this pattern for about three weeks, with the cows stopping their kelp consumption each time they ate ammoniated hay, and binging as soon as they got our hay.

The cycle repeated itself about four times before the snow finally melted enough to go back to the fescue. While I cannot explain, and I'm not sure anybody knows, all the chemical reason why this occurs (other farmers have reported similar experiences) the fact is that we can clearly manipulate the animal's normal cravings by what we feed .

Clearly these animals were starved for minerals. Anytime you see livestock debarking trees and gnawing wooden feeders or licking wooden gates, you can be assured that they are starved for minerals. If yours are exhibiting this behavior even though you are feeding minerals, clearly the minerals are not doing the job. These cows we had ate salty, non-sweet kelp voraciously, letting their sweetened material harden in the box.

Their taste, or desire, was determined by their diet.

If you try kelp, therefore, and have trouble getting the animals to eat it, look around for unnatural protein/nitrogen in your operation. Are you feeding poultry litter? Are you ammoniating hay? Are you using urea lick tanks? These so confound the animal's natural desires that the animal can't appreciate positive alternatives when they are presented.

Another way kelp will fail is to mix it with other things so that the intake of kelp is cut. A cow will only ingest so much salt, or so much magnesium, or so much phosphorus. I appreciate the cafeteria-style approach of placing raw minerals in a multi-compartmented box and letting the animals choose which ones they want. I would suggest using the same principle in kelp feeding. Rather than mixing it in with trace mineral salt or laboratory-concocted mixes, separate the components and see what the cows say. I encourage dairy farmers to just add kelp in a free-choice box, alongside their previously-used mineral, and see which the cows prefer. Typically, the cows will let the old material harden in the box after having a chance to eat good kelp.

We worship today at the altar of the laboratory. No scientist is as smart as the cow. As Allan Nation says, "God gave that cow everything she needs to know to be a successful cow." It's amazing to me how technical or mechanistic we can become in our farming practices, and forget to ask the cow. We shift from square bales to round bales, never asking the cow to choose and let us know which is more proper. We shift from composting to manure lagoons, never asking the cow to choose between slurry-fertilized forage and compost-fertilized forage.

If we would simply ask the cow, many of our uneconomical and unecological practices would be exposed for what they are and promptly replaced with more cow-friendly practices. Instead of subjecting kelp to laboratory analysis and sending it off to some Ph.D., just run your own experiment and "ask the cows." I would even go so far as to suggest that you ask the cows to determine which kelp to use. Many suppliers and types of kelps are available. Instead of paying for an expensive and inappropriate lab analysis when shopping around for the best one, buy a bag of each you are considering, put each in a similar box in the field, and shortly you will notice which one the cows prefer. Use that one; cows won't lie to you. They don't get wined and dined by fat cat salesmen and they have no preconceived notions. They will give you a free, unbiased, God-oriented assessment: bank on it.

The final way to reduce kelp efficacy is to mix everything together (Total Mixed Ration mentality) which will limit the kelp ingestion and therefore the iodine necessary to stop our pinkeye problems. Some of this boils down to faith. Do we really have faith to believe nature can formulate recipes that are as efficacious as those concocted by lab technicians? I for one will cast my lot with nature. It has never failed me.

While I don't advocate eliminating laboratories, at the same time I marvel at how much money farmers send them, and how much advice labs dispense, that is either impractical, does not attack the weak link, or is completely flawed. Too much analytical advice is financed by sales goals.

Kelp should always be fed free-choice, separate from everything else. There is no reason to overfeed the animals — kelp is not cheap — and we don't want to underfeed them either. In the course of a year, consumption varies dramatically. If we feed at a constant level, we are not permitting the cattle to make the choice as to how much they need on a given day. We'll feel it in the pocketbook if we give them too much, and feel it in herd health or performance if we give them too little.

Let's trust our animals to tell us what they need. Listen to your cows, watch your cows, and they will tell you what you want to know.

Chapter 26

Worming

Because I have received so many inquiries about using Shaklee Basic H soap as a cattle wormer, I think it necessary to explain our experiences with this alternative.

First, Shaklee Corporation does ***not*** endorse the use of this soap as a wormer. The fact that many farmers around the country are using it for that purpose does not mean the manufacturer accepts liability in this regard. Neither do I: this is just an explanation of our experiences.

As I understand it, Basic H is really a concentrated protein, made from two soybean enzymes. It is totally biodegradable and has no carriers or harmful additives like most name-brand soaps. You can ingest it in small quantities.

Shaklee uses multilevel marketing, and emphasizes vitamin supplements, health and beauty aids, and other cleansers. I am ***not*** a Shaklee distributor, but there is probably one near you who can get you all the Basic H (stands for Hygiene) you want. It comes in quarts, gallons and 30-gallon drums, and has been used in agricultural applications for probably 20 years.

Many farmers use it as a surfactant (wetting agent) in liquid spray applications. It is the surfactant of choice in much of the biological farming community, when foliar applications of fish emulsion, seaweed or trace minerals are applied. Organic standards committees are divided over whether to allow it as an acceptable product, not because it is feared to be toxic, but because Shaklee refuses to divulge proprietary information. Knowing Shaklee's reputation in the health field, I have no reason to question the integrity of the product.

Our first experience using it as a wormer was about 14 years ago when we had a heifer that would not respond to the veterinarian's recommended wormer. At that time, I received from a friend a multi-page bulletin put together by some Shaklee distributors articulating agricultural and other applications of the soap. It even claimed dairymen could feed it in the parlor concentrate instead of soybean meal to save money and increase butterfat percentages. The only real warning on the product is to avoid eye contact.

I decided to try it on this heifer. She had such a bad infestation of strongyles (liver worms) and was so unresponsive to anything we did that we expected to lose her. We had penned her up in a corral to doctor her. We mixed it in her water for a couple of days, and within a week she completely changed. Her tail was clean instead of hanging with manure. Her coat was slick and shiny instead of dull and rough, and she was kicking up her heels instead of being weak and listless. She was an altogether different calf.

After that we began using it on the whole herd, and have never used anything else for more than 10 years.

We isolate the water and use one tablespoon per five gallons of water. That translates to a little more than a cup per 100 gallons. A cup and a half is fine. That's the nice thing about this: you can't really overdose the animals.

Other folks around the country have used it, and reported effectiveness. By taking before and after fecal samples, efficacy has been proven time and time again. One thing that bothers me is to get a call from someone demanding double blind land grant university studies proving its effectiveness. We must be our own researchers and developers. Anyone who is waiting for government to lead them is going to be way behind.

I am not sure if it will work in a dirty operation. If a livestock enterprise is using conventional unmanaged grazing, allowing access to ponds and riparian areas, housing animals in filth and mud or continuously-used shade trees, then it may well not work. It is not a panacea, and should not be regarded as such.

We try to make sure the cattle drink Basic H water for two days, so that even timid ones will be sure to get it for a full 24-hour period. The cattle readily drink the water, and offer great picture opportunities if you swish the water around so it suds up. They get it all over their noses and faces, looking like a bunch of guys shaving at a tub. It's really hilarious.

An observation I have made over the years is that the soap seems to increase grazing efficiency by about 25 percent. When we put the cattle in their 24-hour paddock, they always eat about 25 percent less when they are drinking the Basic H.

I do not know if the cattle's performance is holding up on less consumption or if they are just eating less. They appear as contented as ever — sometimes more so. I think this would make an excellent research project for someone set up to do it. Clearly, if we can increase forage to beef efficiency by 25 percent, or increase cow-days by 25 percent, that would make continual use of Basic H an option to consider.

As far as I know, parasites do not build up a resistance to it; that too is something to research. We usually worm about half a dozen times throughout the year just to be sure we're getting everything.

The beauty, of course, is that you can worm 100 cows with a couple gallons of Basic H at a cost of a quarter a head. The big payback, though, is that the cattle need not be put through a headgate or corral in order to administer the material. All you have to do is confine the water and let the cows drink. No stress and no fuss.

Many folks have asked about using it as a drench. I guess that goes along with the notion that unless we come in with some shin kicks and broken corral boards we haven't done an honest day's work. Why would anyone want to drench the cattle if they can just administer it through the drinking water? I have no idea if it works as a drench, but I rather doubt it because it is in such low concentrations. It's the time plus the quantity that seems to make it work.

For people who either do not want to use Basic H or for whom it does not work, worming with diatomaceous earth is effective. Most internal parasites are aqueous and the little diatoms, which under an electron microscope look like Tic-Tac-Toe boards with frayed edges, puncture the organisms and rupture them. The cow just excretes them after they die and can no longer hang onto intestinal walls and cilia.

It is generally mixed as a 5-10 percent portion (by weight) of the salt/kelp mineral mix. DE is extremely light, so this percentage will yield a mixture that is nearly half DE by volume. Although retailers encourage livestock producers to use only feed grade DE, I know folks who have used regular swimming pool filter DE and it seems to work fine.

DE can also be used in a dust bag for face flies. The fine powder should not be inhaled, so handle it carefully. It's a good idea to use a dust mask. Whatever spills on the ground or goes through the animal just adds calcium to the soil. So far, most research does not show that the DE harms earthworms and other soil organisms. Here is an example of research that can be easily skewed, depending on who pays the data gatherer.

DE promoters say there is absolutely no negative effects — in fact, it is the only insecticide allowed in grain storage bins that has no toxic level. But pharmaceutical companies have published reports indicating earthworm damage and dung beetle scratching. Of course, they have a vested interest in selling systemic materials — it's a big market.

Let me say just a word about systemic grubicides and wormers. I will not debate the issue of toxicity and just how much comes through the meat. I honestly do not know exactly what, if any, toxicity follows a substance that makes the meat so bad the bugs leave, but I'd rather not play around with such material. I'd rather err on the side of safety and find out I was overly cautious than err on the side of foolhardiness and find out it was deadly.

With conventional materials, several definite things occur. First is resistance. In the very first year of fly eartag use, livestock journals were reporting resistance. When will we ever learn that we simply cannot invent materials fast enough to stay ahead of the adapting insect and microbe world? These bugs can go through several generations of adaptation in a year — sometimes in a week.

Because these materials never kill all the critters they're meant to control, the survivors develop resistance, or immunity, just like you get calluses on your hands when they are exposed to constant pummeling. These survivors then pass this resistance on to offspring and before long virulent strains exist that were unheard of only a year or so before. This has happened in virtually every field of agriculture: weeds, insects, fungus. It is everywhere apparent, and yet the agribusiness community keeps trying to produce the next miracle pill, the next magic bullet. The real tragedy is that so many farmers are willing accomplices in this battle to extract the final farthing from the countryside and deposit it in the urban bank.

The real problem is not the pharmaceutical company; it's the patronizing farm community. If there were no market, the sales of these substances, and their production, would soon cease. Many farmers complain about the alleged "stranglehold" these companies exert on the agricultural community, and yet patronize them every Monday morning with another syringe of this and another bag of that. If this book helps to liberate just a few farmers from this "stranglehold" it will be well worth the effort. Actually, we strangle ourselves. The bondage occurs because of our thinking, not because of someone else's company.

I don't know how many times I've given a lecture and farmers will come up and say: "But I've heard that Ivomectrin is so good. And I've used it: it seems to work."

Well of course they've heard it's good. Many a display ad in farm magazines touts its wonderful characteristics. But why would you believe an ad? And how much university research has it jump-started with seed money?

This leads me to the second big problem I have with all these materials: they really are not necessary. There are much safer materials. We do not have to drive right against the edge of the cliff. We can drive way on the inside of the shelf; there is not need to see how close we can come to the edge. If it were a one lane road and there were no alternatives, then perhaps a little more risk would be in order.

But the industry is asking us to drive on the brink when there are four open lanes on the inside, away from the edge of the cliff. There are enough risks in life without unnecessarily jeopardizing our health and the health of consumers, our environment and our animals.

Finally, these substances are dangerous to handle. They will kill children and pollute water sources if inappropriately disposed. By not having these materials around, I need not worry about my children or someone else's children accidentally guzzling a jug of grubicide, for example. I don't have to worry about some child playing "Old England" with a malathion powdered hairdo.

Part of the beauty in producing clean salad bar beef is that the entire model is people friendly, and that includes the farmers' children as well as the children of guests who may visit the farm and want to see the pretty animals. This is part and parcel of the enjoyment and the people factor of the salad bar beef opportunity.

The bottom line is that there are efficacious remedies for every malady out there, in spite of what the local conventionally-trained veterinarian and the *Merck Veterinary Manual* may pontificate. When they say that "this is the only way to treat" something, take it with a grain of salt. Furthermore, while you may have a serious problem, look further for the cause. How is the grass? How are the genetics?

Remember that Sir Albert Howard, godfather of modern day composting, replicated numerous times the resistance of animals fed compost-fertilized forage compared to animals fed what he called forage fertilized with "artificial manures." At the time, in India, hoof and mouth disease was one of the most feared, virulent, communicable diseases in cattle. He penned carriers with clean animals in adjacent pens, feeding the one group forage fertilized with chemical fertilizers and the other forage fertilized with his aerobic compost from the tea plantations.

The one group consistently contracted the dreaded disease while the individuals in the compost group never did. This type of research is all over the alternative agriculture community, and yet it receives not a mention in the conventional press.

The link between health and disease, between prevention and contraction, is so well documented it's hard to believe anyone is still recommending Ivomectrin or pinkeye vaccines or pour-ons for warbles. We've sold our soul to pharmaceuticals, devoted all our creativity to building crutches, instead of focusing all our energies into building healthy bodies that don't need crutches. It really is that simple philosophically. Our problems stem from incorrect thinking, incorrect paradigms, and not from sick animals.

So what about worming and general sickness? Well, our vet budget is zero. I don't claim to be an expert on cattle sickness because we just don't have any. That's not to say we've never used a vet, but it is so infrequent it doesn't even show up on an annual basis. Instead of putting animals through a headgate and shooting them up or drenching them down or pouring them on, we purchase kelp, Basic H, diatomaceous earth, electric fence, hay shed bedding and clean riparian areas.

Raising livestock is immensely pleasurable when all the animals are healthy and happy. A model that insures that emotional and economic pleasure is worth pursuing: who could disagree?

Chapter 27

Calving

If we will let nature be the instructor instead of our own cleverness, we will look around and find out when our local native herbivore population is having babies. In our area, we look at the deer, and they routinely fawn in April and May, even up until early June. Typically, beef cattle calve in our area in January and February, finishing in early March.

What difference does it make? Why calve when nature does? Here are a number of reasons. First, look at the needs of the calf. When it is born on frozen mud, it has several strikes against it. Ideally, the calf should be born on warm, grassy, dry ground. Generally in January and February and March, it has none of those options. The grass is dormant and wherever the cows are, it has long since been nipped off. The only decent soft bedding material around is some clumps of junk hay the cows wouldn't eat. But they've tromped through it so it's soiled with mud, dung and urine. Great bed.

If it's cold, the ground is frozen. If it's not cold, the ground is soft and muddy. These conditions are perfect for pneumonia, coccidiosis, naval infections and scours. These conditions are not the best for the humans involved either. It's no fun to strip down to the waist and pull a calf in freezing weather.

Within one week, that calf needs roughage to tickle its rumen and get it up and running. Ideally, this material would be highly palatable, almost grain-quality green grass and easily accessible. This calf is on the bottom of the "peck order." Born in late winter, this calf is denied not only green grass, but has a rough time even getting some decent hay. If it gets into a good wad of hay, a bossy cow or bigger calf immediately comes over and butts it away. It's left to fend for itself, and the pickings are pretty slim. About all it can count on is milk.

Clearly this calf is going to be held back from its full potential during these weeks before grass grows. Compare this scenario with the same calf born on a warm, sunny day in April or May. The late winter mud has given way to lush spring grass — grain quality spring grass. The calf hits the ground on a balmy day, has unlimited access to fresh grass, and the temperatures and warm sun sanitize the pasture. Often calves born at this time will actually catch up with and pass calves born in January and February. But there is a mentality out there that those winter-born calves wean off bigger in the fall. It's simply not true. Any farmer who tries the later calving period will not go back to winter calving. It's better for the calves, much more fun for the people (quality of life) and far better for the cow.

Yes, let's talk about the needs of the cow. As soon as that cow calves, her nutritional needs accelerate as well. Not only does she need to produce milk, but she also needs to replace the body weight and energy expended in producing the calf, then begin cycling so she can re-breed. Her nutritional needs are enormous.

She calves in January, February or March. It's cold. It takes a lot of energy to maintain body temperature on a 1,000-pound cow. She needs a pile of calories just to maintain body temperature. Her mineral needs are extremely high. She needs calcium, selenium and others. Hay is notoriously low, especially in vitamins like A and D. Chlorophyll, nature's detoxicant, is unavailable. The grass is brown. She needs the cleansing properties of chlorophyll and its foundation, magnesium, after delivering a calf and having her mammary glands begin functioning.

But all she gets is cold, hay and mud. Is it any wonder the normal recommendation is to feed a few pounds of corn to cows in the winter?. If we just delay the calving period a couple of months, spring grass will give them all the energy they need — for free. The normal recommendation to feed grain or supplement, then, is obsolete as soon as the calving cycle is in sync with nature. It's amazing how creative farmers can be at spending money in the wrong places.

In order to meet all those needs of the freshened cow, the conventional thought is to put more hay to the cows. Of course, that is proper, and the cow will increase her intake during this period by 30 percent or more, but this is all mechanically-harvested and stored feed. If we hold off the calving so that this huge increase in intake will come on lush spring pasture, which is always more than we can utilize anyway, the cow can self-harvest this additional nutrition and it hasn't cost us a cent. Harvesting with the four-legged mowing machine is always more efficient than doing it with a machine.

The reason farmers have had to "get big" to stay in business is that most of us are in the materials-handling business. We haul the feed in and haul the manure out. We turn thousands of tons of soil — move it, fluff it, stir it — and then transport what it grows.

Most folks around here brag about how much hay they make. I like to brag about how much hay I don't make. The average farmer in our neck of the woods feeds hay for 120-150 days per year. We generally feed for only 60-90 days. As soon as we take the materials handling out of the farm, the profit potential becomes size neutral.

A three-cubic-yard front end loader is much cheaper to operate than a one-yard bucket. An 18-wheeler is cheaper to operate per pound than a pickup truck. In the materials-handling business, it's always cheaper by the million, cheaper by the ton. But as soon as the animals self-harvest the forage, there is no inherent economic justification for increasing the volume. This is why virtually every agriculture economics class in America is bogus. The faulty paradigm leads to faulty analysis leads to faulty conclusions.

I listened to an ag econ dude lecturing at a forage meeting about whether it was better to sell hay or buy feeder cattle. He established thresholds of value for corn and barley and stockers that would help the farmer decide what to do. What amazed me was what he had ***not*** considered — the value of the manure by putting the hay through the calves rather than selling it off the farm, the cost of transporting the hay to a new location versus the cost of transporting the calves to where the feed was (it's always cheaper to transport the calves than transport the hay), the difference between the value of well-made hay and poorly-made hay with their concomitant average-daily-gain differences. The list could go on and on, but the upshot was that anyone using his material for decision-making was doomed to be like a horse with blinders on deciding whether or not to jump a high fence when next to it is an open gate.

This kind of analysis has been done routinely in the industry. I well remember listening to the Ph.D.'s touting a new chemical plant growth regulator on the market that retarded the physiological maturation of forages. The research took two fields, ran cattle in both, applied the herbicide to one and nothing to the other. Of course the one that stayed vegetative longer produced more beef per acre which in turn paid for the cost of applying this material.

The research did not have a third field divided into paddocks under a management-intensive grazing regimen. Such a field would have shown the folly of both other options (continuous grazing with nothing or continuous grazing with growth regulator). But I wasn't there to provide the seed money to start that research. Again, thousands of farmers flocked to the chemical store and applied material instead of applying some management to let the animals keep the grass vegetative through good grazing management.

I could fill this book with similar stories, but the point is that farmers must quit operating under the notion that unless they've bounced around on a tractor seat for eight hours and inhaled their daily allotment of diesel fumes they haven't put in an honest day's farm work. Folks always ask me how we get all the work done.

First, we don't have lights on our tractors. Second, we are just orchestra conductors making sure all the animals are in the right place at the right time to capitalize on natural seasonal cycles. Third, the animals do all the work.

By keeping the cow dry through the entire winter period, we

never have to feed a lactating cow with mechanically-harvested feed. Expensive mechanically-harvested and stored feed will go through dry cows ***only***, and they don't need much. This point is critical because it has such a big impact on profit. The lion's share of the expenses in a cow-calf operation is overwintering the dry cow. Depending on what university you visit, this cost runs anywhere from $100-$200 on the Extension service computer printouts. That is a pile of money, and if we can cut it by 30 percent with our calving time, cut it by another 30 percent through our grazing management, we've just taken a big chunk out of the production cost.

It should be abundantly clear now why wild herbivores and ungulates have their babies when they do. It meshes with the seasonal flow of nutrients and weather. It fits.

Just a word about fall calving. Not only is it completely unnatural, but it requires heavy mechanically-harvested feed supplementation to get the same performance as spring calving. It's simply hard on the stock, as well as the farmer's bank account, to operate anti-season.

When we talk about processing, we will discuss this seasonal aspect further, but I think it would be good to note here that spring calving puts these calves at slaughter weight before the second winter. A fall calving approach would require double-wintering these calves and double-summering in order to process them off succulent pastures in early fall. Late spring calving puts the calves right at 18 months a year and one summer later, right on schedule for that ideal harvest time.

Spring calving does shove the breeding season farther into summer, but that is not a problem with management-intensive grazing like it is with continuous grazing. The reason most farmers want all their cows bred by early June is that their pastures begin falling apart around that time. Their calves are huge from having been born in the winter and that is draining the cows' energy (the larger calf wants more milk) much more than if the calves had been born in April and May. This just militates against the deterioration in the pasture quality. To add insult to injury, the heat is playing havoc with the comfort of the cows and the bull, and that further deteriorates summer conception rates.

But if the pastures are kept lush through good grazing management and we have some good heat tolerance selected into the cattle, and the calves are smaller, thereby pulling a little less milk, breedback is not a problem in July and August. Any salad bar beef producer would do well to think seriously about this breeding schedule when he picks replacement heifers. In many cases, it may require a change in seedstock.

But I can guarantee that calving on nature's cycle — whenever the multi-stomached wildlife in your area is having babies — will yield a more profitable enterprise. Performance will be better, sickness will decline, and quality of life will go up for the farmer.

Chapter 28

Weaning

We wean calves in late fall or early winter, depending on the season. If we have a wet fall and plenty of stockpiled forages going into winter, we leave the calves on their mothers as late as possible — even until Christmas.

If it's a dry fall and stockpiled forages are minimal, we wean earlier — even in October — and put the calves on the best hay while the cows stay out and scavenge what they can off the pastures.

Because we calve late in the spring, there is really no rush to get the calves off in the fall in order to rest the cows. Weaning even as late as Christmas, the cows still have at least three months to rest before calving. As soon as we wean, the nutritional requirements of the herd change. The calf needs are high because by this time, cow-calf bonding is more psychological than nutritional. By the time the calf is a hundred days old, most of its feed is coming from the grass, not the cow's milk.

The main change in nutrition, of course, is with the cow. As soon as we take that calf away, we can drop those dry cows to a maintenance ration. As long as the calves are with the cows, we either need to let the calves creep graze ahead of the cows or overfeed the cows. Simply because it's easy, we usually overfeed the cows. In a tight grass situation, like a droughty fall, the sooner we can quit overfeeding the cows, the tighter we can ration out the early winter forage.

By weaning early, we can put the calves on good hay and push the cows real hard on stockpiled forage. The total forage consumed drops, and that is the goal in a tight feed situation. If we have abundant forage because of a wet, late fall, we just leave the calves on their cows longer, overfeed the cows a little, and utilize more of the forage. We want to consume all the stockpiled forages by early January in order to get the manure onto the hay shed bedding pack and to keep the cows off the pastures during the freeze-thaw period.

Furthermore, with our Brahman influence, the cows are not as winter hardy as straight European breeds. That was a sacrifice we made to gain tremendous heat tolerance during the summer. Every decision has trade-offs, and this was one we had to make. Because of our hay shed, we can fudge on winter hardiness. To see those cows perfectly contented and not fighting flies or parasites in the middle of the summer makes this trade-off a good one.

Once the calves are weaned, the two groups of animals do not get back together until spring grazing. By that time the emotional bonding is gone and the cows are having other calves anyway. Occasionally an older calf will pimp off its mother, but that is a genetic deficiency in maternal instincts. Any cow worth her salt will never let last-year's calf nurse to the detriment of her newborn. Rather than ranting and raving at the calf, we need to cull that cow at the next opportunity.

Again, with the Brahman influence, we've had to cull some cows because of this. Those calves are extremely intelligent, and they don't forget what Mama had. You can't blame the calf for wanting some milk. But I do blame the cow for letting the old calf nurse. It takes a strong-willed, aggressive mama cow to put her yearling calf in its place, but a decent cow will do it in a minute.

I owe Burt Smith (one of the few great Ph. D.'s I know) and Bud Williams everything for introducing me to stress-free weaning. I've patched more fences and gates, chased more calves and cows, and lost my salvation more times over weaning than I suppose anything else in my life.

But Burt and Bud's technique is so simple, I don't know why I was too dense to see it myself. The foundational concept is this: when weaning, the more you can keep the same, the easier it is.

In other words, if we're going to wean on Wednesday morning, the more similarities we can keep between Tuesday and Wednesday, the easier everything will go. If the cattle can be in the same field they were in before weaning, that's better. If they can receive the same feed, that's better. If they can see each other, that's better. The point is to minimize the differences between the day before and the day of weaning.

Since we always wean well before we run out of late fall or early winter pasture, we wean out on pasture. We put up a triple electric fence between where the herd is grazing and the next paddock break. No need to put up triple fences along the sides — just the break fence (in between the two groups). The triple wire insures that no calf will go under, over, or through. Normally, of course, a single wire works fine, but weaning is a different ball game.

We bring the herd to the corral and separate the calves and cows. We take the cows back to the paddock they were in when we brought them to the barn. The calves go into the adjacent paddock — the next one in the normal grazing sequence.

Since the calf paddock has ungrazed, succulent grass, the calves immediately begin grazing. They do not bawl, they do not walk the fence. They are completely disinterested in the cows on the other side of the fence. The animals are in the same field, they can see each other, and they are on the same feed. Everything is the same as it was an hour ago except the calves and cows cannot physically get together.

We continue moving the two groups down the pasture, using the three-wire fence as security, for two more days. Until you see it you won't believe it, but the calves do not walk the fence, they do not bawl, they don't do anything. They don't wear a path next to their mothers' paddock. The only bawling is from the mother cows who know the calves are getting better grass. And that bawling is very, very minimal.

It used to be we couldn't sleep for three nights due to the racket coming from the barn and corral. We'd pen the calves up and feed them hay. Everything was different. Their location changed, their ration changed from grass to hay, they couldn't see their mothers: you couldn't invent a more stressful situation. Of course, the sound of splintering boards added harmony to the chorus of bellowing. Such cacophony was wonderful music to sleep by. But those days are gone . . . forever. It truly is remarkable.

In three days, you can take these calves to another pasture and they are as quiet as can be. They do not drift in weight. It is miraculous how easy it is. Again, my thanks to Burt and Bud for introducing me to such a wonderful approach. Like I say, it's amazing how creative we can be at inventing stupid techniques. Now weaning is fun.

Chapter 29

Castrating

For us, castrating is strictly a management-easing technique. I have every reason to believe that young bulls produce excellent beef. I know people who are doing it, and apparently it is fine. I hope more and more will.

But we castrate in order to run everything in one herd. By 90 days of age, a bull calf has secreted enough male hormones to maintain its male vigor through the first year. Rather than castrating at a day or two old when we ear tag the calves, we wait until around 90 days. Even though this puts us in fly season, we've never had a problem.

Rather than banding, which encourages anaerobic infection (tetanus) we cut the tip of the sac off. The seeds and cords are gently pulled out to get as much of the cord as possible until it finally breaks off. Breaking off the cords cauterizes the blood vessels to reduce bleeding. The only knife cut is the tip of the sac.

Leaving the sac tip open encourages air penetration and that stimulates healing. Most infections can't handle aerobic conditions. The best thing you can do for an infected splinter, for example, is to open up the skin and let air get in the wound. The sac can also drain this way. A side cut allows a pocket in the sac bottom to catch the drainage and facilitate infection.

When we're done, we splash Basic H all over the area —

nothing else. The Basic H water will keep flies away for three days, and by that time things should be dry and well on their way to healing. At 90 days the scrotum is not too big, but we have capitalized on bull vigor. It's a compromise to get the best of both worlds. Castrating late, say at 400 or 500 pounds, is a shock to the calf.

If we did not castrate, we could not run the bull calves with the herd the second summer. I'll do almost anything to keep from having two herds. Whatever we lose in pounds gained per animal, we more than make up for in pounds gained per acre because of the larger herd.

We never castrate near weaning time. Castration should be done when the calves are still getting plenty of milk from the cow. The cows will lick the injured scrotum, reducing infection, and the calves can be well comforted by their doting moms. At this age, castration certainly riles up the calf for a little bit, but the stress scarcely lasts for more than a few minutes. It does not set the calf back due to shock, like castrating an older animal, and yet the calf grows almost like a bull. The best of both worlds.

I do not recommend emasculator bands for two reasons. First, the wound time is long. After it cuts off the blood flow and the sac begins to shrivel, the band works its way into the skin and this makes a large wound. Since this takes a couple of weeks, the wound is active for a long time. The longer a wound is active, the more attractive it is to infection, fly infestation, and disease.

Second, this condition sets up anaerobic conditions, which are conducive to tetanus. We are all familiar with puncture wounds and slicing wounds. We like to see a good blood flow coming from a wound because that cleanses the area and we know oxygen can get in there. A puncture wound always sends us looking for our medical records to see if our tetanus shot is till current.

As the constriction proceeds, the outside of the wound ring begins capping over and this shuts off oxygen. Since this occurs over a long time, the risky conditions simply invite tetanus. This is the single reason most veterinarians do not recommend bands.

Many folks like emasculator clamps, but I've heard too many

stories about "one-seed bulls" to put faith in them. Besides, when I pull out both seeds I know I've got them. I can't see anything with the clamps. All I can do is hope I got both cords squashed. I'm a belt and suspenders man, so it's natural I would mistrust something this obscure.

Done correctly as described above castration produces no blood and only causes a few moments' discomfort to the calf. After castrating, the herd should be moved to a good clean pasture to let the calves heal and protect them from picking up infections. This technique is usable for virtually any sized calf.

Steers are more forgiving than bulls on meat palatability. Because a bull has more gristle than a cow, bulls for beef are usually dressed before they are 18 months old. A steer, on the other hand, will produce good quality meat for up to three years. This wide window allows more flexibility for processing and marketing. If we have an abundance of steers one season, for example, or perhaps a late calf, and want to hold one over to the following year, it can still be sold as prime salad bar beef, even though technically it is "old." These animals get huge, of course, but are still high quality.

The size and higher total sale price does not compensate for the second winter, but the marketing flexibility is worth the growth efficiency sacrifice. One year we might have a higher percentage of heifers that we want to keep or sell as breeding stock; having a couple big steers carried over from the previous year helps level out our marketable inventory.

Chapter 30

The Eggmobile

Using nature as a guide, we notice that birds always follow herbivores.

The egret perched on the rhino's nose is there in symbiosis with the herbivore. He picks out parasites, scratches through dung, and acts as a biological sanitizer for the big brute.

To mimic this natural principle on a domestic scale, we follow the cattle in their rotation with a portable chicken house we call an "eggmobile." It is nothing fancy — a trailer 12 feet wide, 20 feet long and about 5 feet high. Laying hens scratch through the cow paddies, eat out the fly larvae and parasites and spread out the dung to stimulate nutrient cycling.

Instead of putting the cattle through the headgate and pumping them full of systemic pharmaceuticals, we just pull the eggmobile around behind them and let the chickens do the work. It's similar to the pigaerators in that the chickens do not require minimum wage or unemployment compensation. Instead of spending money on wormers, we have a by-product: several thousand dollars worth of eggs a year.

The farm economics change completely when we convert expenses into incomes.

The eggmobile has a small trap door to let the chickens go in and out, along with a ramp. The floor is 1 inch X 2 inch hardware

Photo 30-1. *The eggmobile, a portable layer hen house, follows the cattle as a biological pasture sanitizer. By-product is $3000 to $4000 worth of eggs per year.*

cloth so the droppings fall onto the pasture. A regular trailer jack and clevis hitch on the front make it easy to hook up to a tractor or pickup.

Nest boxes along the sides like saddlebags allow outside gathering. The boxes are sectioned into one-foot nests. A total of 32 give plenty of boxes for up to 200 hens.

Every 100 hens eat seven pounds of proteinaceous critters per day: worms, caterpillars, grasshoppers, crickets, beetles, grubs and larvae. It takes roughly 100 hens to clean up after 100 cows and the birds will cover roughly two acres per day. We try to move the eggmobile every other day and keep it within four days of the cows because the fly cycle is four days. If we wait longer than that, flies have already hatched.

As long as they are kept moving onto fresh ground, the birds pick up 70 percent of their diet from the range. That means we can produce eggs for 25 cents a dozen. Because they are of such superior quality, we can get 50 percent above regular market price. This gives us a nice gross margin.

When we want to move the birds, we just make sure the trap door is closed on them the evening before. Then we can go out in the morning and move them wherever we want to. This is a land extensive system. I do not recommend the eggmobile for grazing systems under 50 acres. In such small acreages, the birds get too familiar with certain areas and keep homing back to them. For example, they will come back to a barn, a garage, a garden — the back porch.

The acreage needs to be big enough that the birds only know the eggmobile as home. Then they will always come back to it at night. In the small acreage, the birds will stray too far away to find their way back by nightfall. Forestal areas to break up the fields also helps keep the birds from staying too far. One of the keys to keeping the birds from homing in on a permanent fixture is to move the eggmobile often. The more often it is moved, the less tendency the birds have to home in on anything else. If it sits in one place for a week or more, the birds will begin laying eggs in the grass and fencerow, and will wander too far away.

Generally the birds will go 200 yards away from the eggmobile in a day. They get used to the routine, and of course it has water and grain for them. We use several feeders and put in whole grains, cafeteria style. We give them whole corn, a small grain, meat and bone meal, ground feed grade limestone and oyster shell. They fluctuate their intake of the different items throughout the year. If we have a heavy infestation of grasshoppers and crickets during a drought, for example, the birds scarcely touch the meat and bone meal.

We use only non-hybrids for the layers because they have more brains than hybrids. Believe me, a chicken needs all the brains she can get. Old traditional American varieties like Barred Rocks, Rhode Island Reds, White Rocks and New Hampshires work quite well. If they see a hawk overhead, they run under the eggmobile and wait for the danger to pass. I've watched this happen numerous times. The birds need to be aggressive foragers and scratchers, and yet be domestic enough to not run off in the woods and leave.

Some wild birds like game chickens or bantams have a tendency to be so wild they don't stay around. When the birds stay close by the eggmobile, which is almost always close to the cows, predators are not a real problem. We expect to lose a few birds a year just as a cost of doing business. There's no free lunch. But if the birds were religiously closed in every evening, I don't think we'd lose any to predators.

An Amishman from Holmes County, Ohio built an eggmobile and got tired of going out in the evening to drop the trap door. He fixed a hasp on it with a peg and threaded a string through the peg. He attached the other end to a windup alarm clock, which he fastened to the wall of the eggmobile. He set the clock to go off at dark — chickens go to bed real early — and when the alarm sounded, the windup gizmo would spin and wind up the string, pulling the trigger out of the hasp. What a clever idea. These are the kind of refinements I hope people will keep adding to our basic concepts.

Based on what I've seen from a pasture sanitation perspective, I'm convinced the eggmobile would be worth having even if the chickens never laid an egg. In fact, if you just stocked it with a

Photo 30-2. *Laying hens from eggmobile scratch through cow paddies, eating fly larvae and spreading out the dung for better nutrient cycling. "Letting animals do the work" is not only fun, but also extremely economical.*

bunch of cheap old ornery roosters and by the end of the season foxes ate the last one, it would still be worth having.

They absolutely destroy cow pies. In fact, walking through the paddocks it's difficult to find where the cow patties were. The incorporation into the soil is that complete. And they will completely debug the farm. In a hot, dry period when the grasshoppers and crickets jump up in front of you when you walk, like water breaking in front of a boat bow, the chickens will clean them out.

Since the birds go under the electric fence, they will take care of a couple paddocks at a time. We try to move the eggmobile every two days, generally parking it on an old cross-fence site so the birds are right between two paddocks. They do not tear up fresh cowpies. They only tear them up when they have critters in them — after a couple of days. Chickens prefer their cowpies properly aged.

In the winter, we empty the eggmobile and park it until spring. It's too airy for winter and we find that the birds do not go after winter-dropped manure anyway — no bugs. As a result, any winter benefits are minimal. But at the first sign of flies in the spring, we run the eggmobile out to the paddocks where patties are thick following stockpiled fescue grazing. In short order the birds clean up the area and we have not had to do any pasture dragging with heavy metal.

Training the birds to go in at night can be a challenge, but they will learn. We only use adult birds in the eggmobile. Never put in birds that aren't laying — they don't have enough reason to go in. But the nesting instinct is strong enough to make them go back in when they are outside. When we first put a new batch in, we lock them up for three or four days to get them used to their new home, where to find water, feed, and where to lay eggs. Then the first day we let them out first thing in the morning.

They must go back in throughout the day to eat grain and lay eggs, so by evening they've gone up and down the ramp several times. The first night, many will not go in. We go out after dark and put the learning disabled ones back in (the LD birds).

After a couple of nights, we go out with a long stick and sit under the eggmobile. If the birds cannot come under it for the night, the only other place to go is out in the field, and they do not want to do that. This technique works extremely well because the birds are forced to go up the ramp on their own volition. Within a couple more days, they will all learn. Once they are trained to go in at night, they never stay out. They truly are creatures of habit.

The eggmobile can be any portable henhouse. The first one I made was 48 square feet and mounted on a three-point hitch. I've seen people build them on hay wagons (be sure to brace it so it won't flex). They can be flat-topped, long, low, high, square, you name it. Let your imagination run.

The point is that the birds do a much more effective job of sanitizing paddocks than anything human or mechanical and they love to do it. As a by-product, you get the world's most wonderful eggs and they can be sold for cash or bartered for wonderful things. We give eggs to the veterinarian and usually do not receive a bill. We give eggs to our neighbor who owns the big cattle truck and he hauls our cattle to the slaughterhouse in the fall. It's the cheapest cattle hauling I know.

We give eggs to another neighbor in trade for "big tractor" work when we need it occasionally. We've given eggs to doctors. Our mechanic often gets eggs. We trade them for welding wire spools, remember. The list is endless. Few things brighten up a person's day than getting a dozen of real honest-to-goodness farm fresh eggs. You always get more than the eggs are worth. And it builds up great goodwill in the community.

Birds and herbivores are a wonderful fit, and by incorporating them symbiotically, truly the whole is worth more than the sum of the parts.

Chapter 31

Soil Fertility

Just for a moment, envision the most fertile area on your farm or ranch. Don't pick an extremely unique place, like right next to a river. Rather, think of that knoll or small area that consistently produces the highest amount of plant material per square foot. It may be a wooded area or a grassy area, but each farm has such a place. What if your whole place were that fertile?

I've often looked at my favorite spots and daydreamed about what it would be like to have the whole place like that. One spot, near the equipment shed, would probably yield 1,000 cow-days per acre per year. Although for certain reasons it may be impossible (soil depth, aspect, etc.) to totally duplicate this performance farm-wide, such dreams are the things that encourage and challenge us.

Go to that area with a shovel and do a little digging. Smell the soil. Feel it. Lift it. Let it fall through your fingers. Look for little tiny critters, like worms, beetles and centipedes. Now go to an area that is poor. Repeat the procedure. I know agronomists have all sorts of fancy names for soils depending on their parent material. I know that soil testing laboratories establish cation exchange capacity, which measures soil potential.

But I've found that regardless of these scientific measurements and designations, fertility principles span all climates and soil types. In fact, if you don't let the scientists limit you too much, you can totally change the productivity of that poor soil to nearly that of the favorite soil spot given enough time, the right materials and an indomitable spirit.

In his classic, *Plowman's Folly*, and then in his second treatise, *A Second Look*, Ed Faulkner detailed the principle of fertility development, from the soil up. Certainly his little bombshell, *Soil Development,* turned many a head before it was relegated to the dusty back shelves of agronomy libraries. He debunks much of what modern agronomists consider soil gospel, and articulates perhaps better than any scientist how fertility can be built up without expensive off-farm inputs. I agree with him.

Much of our farm 25 years ago resembled those Soil Conservation Service pictures in booklets describing the tragedy of American soil erosion. Many gullies measured 10 feed deep and one measured 14 feet. Most of the land had no black topsoil. The red clay came right to the surface. Underneath, shale came to the surface in varying degrees of closeness, often rising clear to the surface in large bare spots. We still have trouble putting in temporary electric fence stakes in some places.

On one of these hillsides about 15 years ago, we were spreading some barn manure on a shaley part when all of a sudden the spreader advance chain broke. Nothing to do but shovel the manure out. Of course, we flung it as far as we could, but right in that spot, we put on an extremely liberal application. That 100 square foot spot has never received any special treatment since. Still, it grows five or six times the forage of the area adjacent to it, greens up earlier in the spring, stays green later in the fall and basically doesn't resemble the surrounding ground at all. The soil is black and loamy, with abundant soil critters. It takes drought better and grows back much faster when mowed or grazed.

Another fascinating thing about this spot is that whenever we graze that two or three acre shaley paddock, that 100 square feet gets eaten into the ground in the first 10 minutes. Furthermore, the cattle tend to lounge on that spot, even though it is not on the top of the hill or in the same plane as their normal lounge spots on that field.

Now, if only the rest of that area could be made just like that spot. That is my dream. The exciting part about that dream is that I know the rest of the area can be made like that spot and I know exactly what it would take to get it there — several broken manure spreaders!

We have another field that faces east and was always poor. One year in the early 1960s we had a couple loads of rain-spoiled hay that we mixed with some manure to build two compost piles, one on either side of the hay wagon. We fully intended to spread them the following spring, but time got away and we never did spread it. Here again, many, many years later, I can show you those spots, the cows can show you those spots, and the earthworms can show you those spots. They are drastically different from the surrounding area.

Soil fertility is not some elusive, nebulous concept that we can never reach. I guarantee that whether your soil is sandy, rocky, loamy or shaley, if you had done the same thing on two spots that we did, the response would be equally as dramatic, and equally as long lasting. True fertility does not come and go with the season. It can't be bought in a bag.

True fertility comes about as many different factors interact to boost soil organic matter and biological activity. Soil scientists often describe the positive productivity of "new ground." The pioneers certainly benefited from new ground. Row crops following forages always produce better than when they follow other row crops or each other. Crop specialists don't have answers to this, except to label it the principle of new ground.

What was your farm like before settlers came? What was your new ground like? Probably it was like that fertile spot you just visited. Whole cultures have risen and fallen because they couldn't figure out how to make all the area like the good spot, and instead made all the good spots like the bad ones. Organic matter is the key.

When dead plant material decomposes it gives off carbon dioxide. When this gas contacts moisture, an acid is formed In a chemistry lab, if I want to break out the minerals in a rock, I can use many acid reagents: sulfuric, hydrochloric, etc. But the most efficient one is carbonic acid. That is the one formed when carbon dioxide meets water. Yes, it's the one formed in the soil as organic matter decomposes into humus.

Most of the time our low fertility problems are not mineral problems per se, but a lack of carbonic acid production. It is the carbonic acid that breaks the minerals out of the soil. Without an active decomposition program, carbonic acid is not produced and the entire system shuts down. It is common to measure calcium, for example, and find a small part available but a huge storehouse labeled "unavailable." Is the weak link calcium? No. The weak link is carbonic acid production to access it. Faulkner eloquently defends the notion that rocks in the soil have as many minerals in them as they did centuries ago. The only problem is that we've destroyed the ability of the soil to access these minerals.

Whatever you put on the soil that burns your skin, or is hazardous to you will also burn up earthworms and soil microbes. In one handful of healthy soil, there are more living organisms than there are people on the face of the earth. How dare we treat the soil like dirt?

One pound of organic matter holds four pounds of water. Not only does it allow water to penetrate faster, but the total holding capacity is greatly increased when organic matter is higher. This has everything to do with floods, erosion, drought tolerance and capillary action, as well as aeration, root penetration and cation exchange capacity.

Weeds describe soil fertility. Rather than attacking weeds, they need to be seen as symptomatic of fertility or management problems. They can help diagnose problems. When organic matter is where it should be the soil adjusts itself, the carbonic acid breaks out the minerals necessary for proper balances, and true fertility occurs.

Certainly remineralizing the earth can speed up this process. I cannot disparage the folks like John Hamaker whose research shows the value of remineralizing the soil. But studying nature reveals that first the organic matter must begin cycling, and then other things can be added. Every farmer must have an echelon of priorities. The more esoteric refinements must be delayed until the organic matter is given its proper place.

We have fields that at one time were highly acidic and full of species that indicated low calcium. Today, those same fields are full of clover, which indicates heavy calcium availability, and those weeds are gone. We never spread minerals, but applied compost and pulsed the perennial sod through managed grazing. We probably could have gotten where we are faster had we applied minerals, but at the time we did not have the money nor the graziering skill to capitalize on better grass anyway.

The point I'm trying to make is that we can buy time with cash or we can let time be our friend in many of these fertility endeavors. It depends on our goals, our time schedule, and our bank account. But for sure, expensive off-farm items are ***not necessary***. That is the point of this discussion.

Livestock fed forage from poverty soils will still excrete dung rich in nutrients that differs only slightly from dung excreted from animals fed forages from an extremely fertile soil. The enzymes and body chemistry of living organisms can seemingly create necessary elements out of nothing. Students of such things call this process transmutation. Most scientists don't believe this occurs. I don't want to quibble over terminology. All I know is that you can start with poverty soil, and without adding anything but careful management and nutrient cycling, create a soil that does not resemble in appearance, structure or forage productivity the original impoverished material.

The only kind of material to apply to the soil is what nature applies. Plant residues and animal excrement, along with mineral powders that blow in on the wind: these are the things that have built soil for centuries. They still work if we will not be so bound and determined to destroy them through inappropriate tillage and chemicals.

As these build up organic matter, earthworms begin hatching

and add their action to the soil. Earthworm castings have 1.5 times the calcium, 3 times the magnesium, 5 times the nitrogen, 7 times the phosphorus and 11 times the potash of the soil through which the worms dug. And 12 worms per square foot will yield three inches of soil per year. Andy Lee in his recent book, *Chicken Tractor*, makes the enlightening observation that soil does not deepen, it 'uppens.' That is a perfect way to look at what happens in the soil.

It also should give us pause to think about what is happening when rocks begin appearing in fields and we see the soil shade turn from black to brown to red to white. The bottom is not coming out, the top is leaving.

I've attended agronomy lectures where the Ph. D. stands up and says: "The three biggest elements in the soil are carbon, hydrogen and oxygen, but we're not going to talk about them. We're going to talk about nitrogen, phosphorus and potassium." Isn't it amazing? If we put the kind of attention on the first three that we should, everything else would fall into place. And by inverting the priority in our minds, we have destroyed the first three, which are the foundation for the proper retention and working of everything else. It's hard to believe how creative we can be at lousing things up.

You need not have an agronomy degree to know what a productive soil looks like on your farm. The key to soil fertility is to duplicate that appearance on every other square foot. It's really that simple. And whatever is consistent with nature's principles, and is within the confines of your resources, do it. Fertility is not something we "buy in" as much as it is something we "build in" from within the farm organism, as it were. And when it is there, it energizes itself and continues getting better with a minimum of outside input except good management. The soil is extremely resilient, fortunately for us all, and will respond to good treatment better than we ever thought possible. May we all dedicate ourselves afresh to quitting fertility-debilitating practices and implementing fertility-enhancing practices.

Chapter 32

When to Apply Soil Amendments

First, we only want to apply material to biologically active soil. In his wonderful book, *Fertility Pastures and Cover Crops*, Newman Turner explained the folly of soil sampling. Depending on whether the sample was taken during the dormant season or spring flush, drought or wet conditions, the nutrient readings can differ by more than 200 percent. We should never apply nutrients in any form to cold soils.

Going one step further, we should limit our application to soil that is actively growing something. That is why side-dressing in row crops has become common. Fast-release acidulated fertilizer was giving the soil more nutrients than the corn plants could metabolize in their infancy and there was a dramatic shortfall in nutrients at silking time. The corn was overindulging on Christmas dinner as a toddler and starving to death at the height of its virility.

With management-intensive grazing, we can assure a vigorously growing forage numerous times throughout the season. We want to apply nutrients at those times. But we have one other concern: the forage growth curve.

Voisin advised applying nutrients prior to normal forage slumps in order to raise the valleys and, by withholding amendments, likewise reduce the peaks. The result was a more constant forage volume. The spring lush, for example, grows plenty of forage anyway. It doesn't make much sense to apply nutrients that will only increase the oversupply of spring forage.

By the same token, if we wait too late in the summer, we stand a good chance of hitting a drought, and the extremely slow growth reduces nutrient assimilation. We want neither extreme.

Ideally nutrients should be applied between spring lush and the summer slump. That time meshes well with composted shed bedding. Grazing will normally begin in early April, and the composting process begins immediately thereafter. After hay harvest in early June, compost can be applied to stimulate forage regrowth prior to summer dry periods. That allows a 60-day composting period, which is almost perfect for nutrient stabilization.

After being mowed for hay, the grass is hungry, wanting to enter the fast growth period, and ready to metabolize nutrients. By applying compost at that time, we can artificially simulate May conditions again through the flush of nutrients and biologically-active material. It mirrors spring wake-up and can go a long way toward reducing the summer slump period.

The second time period to apply nutrients is just prior to the fall jump in forage production. Generally we are coming off a dry period and wanting to capitalize on the first significant rain following September 1. When a rain is in the forecast, we like to spread compost. This application stimulates growth for winter stockpiling, helping the fall forage flush look more like the spring flush.

William Albrecht, renowned agronomist from Missouri, said that we should apply amendments when nature does: following spring flush for grass, when dead tops fall over and add organic matter, and in the fall when the leaves begin to fall. We have always found our most long-lived fertilizer benefits from applications at these times.

Lawn care specialists always fertilize in the fall. It increases root mass and adds a lot of energy for overwintering food supplies. Animals always eat extra in the fall, adding body fat to get them through the winter. The pasture is no different.

Only nutrients that feed the decay cycle should be applied. Anything else is either counterproductive or inappropriate (waste of money). Acidulated chemical fertilizers are toxic to the decay cycle, as is slurry spread from liquid lagoons. When that material contacts earthworms, they are as dead as if they had been hit with DDT. These highly soluble, acidic materials not only destroy soil organic matter, but they require large amounts of calcium to buffer and correct the problems. It's kind of like pharmaceuticals that create more dangers from side effects than the dangers from the disease they were supposed to cure.

Composting, the judicious decay of organic materials, mirrors nature's fertility cycle. To implement it on our farms pays big dividends. We can enhance it through good management and labor. Most of our soils can use a little tender loving care. They've received something else for far too long.

Ideally all soil amendments would be applied right ahead of a rain that is just enough to wash them in but not enough to have runoff. Cloudy days are better than sunny days. One advantage of compost is that because it is so stable, it can be left in a pile until conditions are right for spreading. Furthermore, if we happen to miss good spreading conditions, it will not lose nutrients as rapidly as raw materials or chemical fertilizers.

Timing the applications can make a world of difference in how efficient our amendments are utilized. This timing must follow the season, in a long-range way, but also the day's weather, in a short-range way. The differences in the way the soil assimilates nutrients justifies the extra planning effort.

Chapter 33

How Much Hay?

Should we try to eliminate haymaking? Controlled grazing certainly reduces the need for hay, but should elimination be a goal?

Certainly the answer depends on climate. Year-round forage growth reduces winter hay feeding. The colder the climate, the greater the hay requirement. In extremely dry cold areas, swathing is practiced extensively. This is like freeze drying. Accumulated forage is swathed in the fall and then the cattle are offered one windrow a day across the field, portioned out with electric fence. In wet areas, however, this technique will not work.

The drier the climate, the longer the window of palatability for accumulated forages. Whether they are still on the stem or cut and left on the ground, as in swathing, brittle environments offer a long window of palatability because moisture is the single largest forage-deteriorating factor. Nutrient leaching is extremely high in non-brittle environments. In a brittle environment, standing grass can still be nutritious and palatable 200 or more days after initial grazing acceptability, but in a non-brittle environment this window of palatability is often less than 60 days. This means if it is not utilized, it is lost. Plenty of temperate area livestock producers force their animals to eat cardboard and compensate with grains, silage and expensive concentrates, not to mention pharmaceuticals.

These kinds of things illustrate the fact that there is no grazing paradise this side of eternity. In brittle environments, water is scarce and volume is low, but strength (Brix index and mineralization) is high. In temperate areas, volume is high but quality per pound is lower. In hot areas year-round green vegetation is available and forage quality is generally good but parasites are high. In cold areas forage quality is perhaps a little better and parasites are low — but there is no growth in the winter.

Under optimum conditions, what are worthy goals for acreages allotted to hay, stockpiled winter feeding and seasonal grazing?

A no hay enterprise is nearly impossible in North America because of the heavy spring forage growth. Any forage growth chart will show the greatest volumes occur in spring. Certainly warm season, drought tolerant species can help fill in the summer slowdown period, but still the most volume will be produced in the spring. And unless there is an unlimited stocking flexibility, this production surge cannot be grazed when it is most palatable.

In order to push productivity past the 400 cow-day level, the forages must be harvested before they become too mature in order to make room for new growth. Otherwise, not only will the forages become unpalatable, but regrowth will be delayed and then total forage productivity will be stunted. It is imperative to maintain all the forage in a growing state during the growing season.

To help visualize the potentials, I've developed some guidelines that illustrate the necessity for haymaking and how much acreage can be budgeted for that purpose under several different scenarios.

Dates	% Growth	% Time	% Acres Hay	/ Graze
Set 1				
Apr. 1-Jun. 15	33	20	26	74
Jun.15-Oct. 15	33	33		
Oct.15-Feb.15	33	33		
Feb.15-Apr. 1	0	13		
Set 2				
Apr.1-Jun.15	40	20	50	50
Jun.15-Oct. 1	30	30		
Oct.1-Jan. 15	30	30		
Jan.15-Apr. 1	0	20		
Set 3				
Apr.15-Jun. 15	50	20	60	40
Jun.15-Sept. 1	20	20		
Sept. 1-Nov. 1	30	16		
Nov. 1-Jan. 1	0	16		
Jan. 1-Apr. 15	0	30		

The first set illustrates the forage growth in a southern region, where the growing season is spread long and drought tolerant species are more the norm. This shows minimal hay requirements.

The second set of figures illustrates forage growth in our part of the country. The peak occurs in May and tapers off into June, with a hefty summer decline and then a spurt in the fall after the first rains come. The final set of figures illustrates an extremely cold climate where the grazing window is shorter in the winter, requiring more stored forage. But the huge spring lush is also larger in these areas.

The point of all this is not to eliminate hay, but rather to use hay as a tool to stimulate forage production and minimize stagnant growth inefficiencies. Perhaps in our mad rush to eliminate hay, we will eliminate 100 cow-days per acre by not efficiently capturing seasonal growth spurts.

If we fail to utilize the spring lush, its window of palatability closes and we cannot capture that forage. If we harvest it, however, and store it, we can not only use it during the dormant season but perhaps more importantly, will stimulate that acreage to convert more solar energy into grass. We will also ensure that those acres remain vegetative and nutritious farther on into the season.

I must point out that under conventional continuous grazing the numbers are far different. For example, here in Virginia it is common to hay only 20-30 percent of the total farm acreage and graze the rest year round. Because the annual grazing cow-days per acre averages about 70, the total farm's carrying capacity is so low that a few acres of hay will carry the herd during the dormant season. These farms get 200 or more cow-days off their hay ground by letting the grass express itself and harvesting mechanically. It is by far the most productive forage acreage on the farm. We, on the other hand, never get as much total production off the acreage we mow for hay as we do off the acreage that is only grazed. The reason is that hay is allowed to go past that second break point on the growth curve. Yes, it makes more volume, but it takes much more time to put on those extra inches. Meanwhile, where we're grazing (shearing) on time, the grass is staying in that blaze of growth period.

Rather than being scared off by the prospect of mowing half the acreage for hay, farmers should accept the challenge of increasing the farm's carrying capacity through controlled grazing to such an extent that double or triple the amount of hay will be required to carry the livestock through the dormant season because the number of animals escalates. This vision helps us see through the sweat streaming down our faces while we make hay.

These charts also help clarify selling strategies. We slaughter beeves in October or early November, following the fall lush. Our lowest cattle numbers are in the winter. This certainly helps to reduce the hay and stockpiling required. A limitless stocking flexibility, with the option of liquidating in the winter, would certainly change these figures too. But that is quite uncommon. The numerically constant herd is far more typical, but far more challenging in view of the tremendous seasonal fluctuations in forage growth.

My grandfather-in-law always said: "You can never have too much hay in the barn." I've seen the wisdom in that statement many times. Our barn full of hay is like a big safety valve. It lets us see when we can expand our carrying capacity and when we are exceeding it, long before we get to a crisis.

Hay is one of many tools that the grazier uses to help match forage to cows, to adjust availability as growth fluctuates. It can be viewed as an extended pasture, a part of the grazing system.

Making a large volume of hay is not our goal. Our haymaking goal is merely to even out the forage growth and give additional flexibility to our planning.

Chapter 34

Making Good Hay

Too often we view making hay as a mechanical process. But it is far more than that. Making good hay is an art as much as it is a science.

In all the articles I've read about making hay, the one element I've never seen is the same one missing from most research: nobody asks the cows. Why is it so difficult for us to believe that God gave the cow everything she needs to know to be a successful cow? Certainly that has a comical ring to it, but there is a serious undertone to the question. Why is it that whenever we want to know something we assume we can ask a bunch of white-coated guys in a laboratory peering through complicated machinery and expect them to come up with the right answer?

For example, take chemical drying agents on hay. Chart after chart shows the quicker drydown of the forage, but has anyone dared to put hay dried with acid and hay dried without in front of the cows, cafeteria style, and see which they prefer?

Forage breeders are producing all kinds of nifty plants and putting them on the market, from spreading alfalfa to imported this or that. Who is comparing these forages to what is growing in the road side ditches, offering the cows a taste test? Certainly not all new introduced forages are poor, but just because they produce more tonnage, or are resistant to this or that, or can uptake more of this or that, does not mean they are the preferred feed for the cattle.

Remember that salad bar — cattle like their variety, too. They prefer eating a variety just like you and I do. It's not only more nutritious, it's also more delectable. Polycultures are always more environmentally sound, all the way around, than are monocultures.

The whole issue of haymaking as an art came home to me a few years ago when a farmer called me for some advice: he couldn't get his stocker calves to eat enough hay. He was having to buy corn and soybean meal to supplement their hay because these 500-pound calves would only eat about 8 or 9 pounds of hay a day.

At the very same time, our identically-aged and sized calves were consuming 20 pounds per head per day. In fact, at about that time I read an Extension bulletin that said fiveweight calves should eat 12 pounds of hay a day — that was a goal. I scratched my head and wondered what was going on. When I looked at the hay the farmer was feeding, it was no wonder the calves balked. It was what I call bedding hay. It was certainly not fit to eat, especially for calves.

I realized then just how much difference there is in hay. To be sure, much of what is written about hay is helpful. In fact, the horse industry has probably done more to promote good quality hay than any other single agricultural enterprise. I think one of the reasons is because almost all horse owners buy their hay; they don't have to make it. They can afford to put more attention on watching what the horses like because they are putting little attention on making it. It's another one of those cases where everyone has just so much creative observational energy, and if it all gets used up one place there's nothing left for someplace else.

Anyway, there is no question that if cattle farmers fed their calves with horse hay, they would never have to supplement with grain. One thing you never see: horses being fed round bales.

Traditionally, farmers scythed hay and forked it onto a wagon by hand. Until the baler, hay went into the barn in fairly noncompacted form. Old barns, constructed with vents on and near the roof, allowed the hay to vent moisture because farmers knew hay would continue to lose moisture as it cured in the barn.

To further preserve the solar dried forages, farmers would salt it down as each layer went into the barn. This not only helped draw out moisture, but it also increased palatability, much as does seasoning food. Along came the hayloader, that wonderful invention of an inclined plane that picked up hay in the windrow and pushed it up nearly 10 feet, where it fell onto the wagon. That is the way we made hay until we began square baling after 1980.

The loose hay did not pack nearly as tight as baled hay. We could put up extremely tough hay, salt it down on the wagon, and put it in the barn without it molding. The long forage stems, running every which way, encouraged venting holes to let the hay "breathe."

With the advent of the square baler, many farmers stacked hay on edge in the barn. The rationale was that this kept it from packing too tight because the weight of the pile pushed into the stems in the bale, rather than across them. Many farmers still use this method, saying that the hay cures better with the stems running vertically to wick the moisture out the top of the pile. It is certainly harder to stack a pile of hay using this method, but the fact that so many farmers, at least until recently, went to this extra effort to prevent mold and encourage good curing illustrates that haymaking is more than mechanics. It is an art form.

Enter the age of the round baler. This machine puts all the stems one way, locked inside the bale, and not protruding to the outside where they can wick out moisture. Furthermore, this machine wraps the hay so tightly that it absolutely can't breathe.

Perhaps round bales exist that are not moldy inside, but they are few and far between. Stick your nose into one and sniff deeply. The round baler exacerbates this problem because it can accept wet hay, whereas the square baler won't.

I've seen square bales fed adjacent to a stack of round bales, and the cows will not touch the round bales until they get good and hungry. This would be a wonderful test for farmers to do. Round bale a windrow and square bale an adjacent windrow. Put both under roof the same day and feed out the same day, letting the cows choose which one they want. If I were a betting man, I'd put money on the square bale.

There is no reason why livestock should need grain in the winter. We make sure our stocker calves get the best hay and feed the junky hay (rained-on, tough) to the cows. No one makes perfect hay every time, and that is one of the advantages of having a brood cow herd: they can utilize the junk. Much of the profit in any business is geared to minimizing losses. If we make some bad hay, to feed it to calves simply compounds the error. The calves go backwards, we might have vet bills, and probably will need to feed some grain.

But with brood cows, we can turn that poor hay into an asset, and not waste good hay on them. That way we can feed only our top 25 percent to the calves and both groups win.

Although the calves will not come through the winter looking like grain-fed calves, they will make compensatory gain in the spring. The ability of a calf to gain weight is in direct inverse proportion to the prior 100-day gain. In other words, a calf gaining 2 pounds a day during the winter will generally drop to a pound or less when it goes out on spring pasture. A calf gaining only half a pound a day during the winter, however, will often gain up to 4 pounds a day when turned out on lush spring pasture.

That is why thin-type calves often bring more per pound at the sale barn in the spring than fat calves. Spring compensatory gain is the cheapest weight you can ever put on a calf. Letting calves come through the winter a little on the thin side, but with a regular frame, is far cheaper than trying to make them gain during that period. It's like trying to make a cow lactate during that period. A hundred days after beginning spring grazing, there isn't enough difference between the wintered fat calf and the wintered thin calf to even talk about. Keeping off the butterball fat will not reduce frame development. Going into spring, the fat and thin calf will both have the same frame; the only difference is the amount of flesh hanging on it. And that flesh can either be put on early with expensive supplemental feeding or later with cheap spring forage. The only difference is a few days and a pile of dollars.

Hay should be mowed late in the day because that is when the Brix is highest. Brix measures the sugar, and plants can vary dramatically in Brix value. Not only does it go up as the plant reaches maturity — that accounts for the sweetness in mature fruit, for example — but it fluctuates every 24 hours. The Brix index, as well as the energy of the plant, drops during the night because the plant lives off stored energy. When the sun comes up photosynthesis kicks in and the energy and Brix come back up.

The same cycle occurs right in front of falling weather (when a plummeting barometer indicates rain is coming). To protect itself, the plant sends its energies into the roots if a thunderstorm is coming. The move apparently follows the drop in barometric pressure.

The secret to making square bales is a lot of wagons. We can put 500 bales on wagons in an afternoon, push them under roof and unload them early in the morning when it is cool. Square bales are much less arduous when we work in the barn only in the cool of the morning. When we finish, the dew is off and we can rake some more and bale after lunch. The key is to not have to stop and unload wagons in the heat of the afternoon. If you can get several hundred on rubber, you can make good hay without killing yourself.

A lot of farmers agonize over the work of square bales, but if the extra labor eliminates the need to feed any animals grain because the hay quality is better, the bottom line is still better. If junk hay is worth something, then certainly good hay is worth far more. The problem is that when we view haymaking as just a mechanical procedure, there is no justification for putting extra effort in making good hay. I suggest that this is another case where some extra care pays for itself.

We salt down the hay as it goes in the barn, sprinkling about 25 pounds per ton. That helps it to cure and definitely increases the palatability. It draws moisture and keeps the stack from heating as much.

We do not plant hay, but rather make hay on paddocks that have gotten away from us in the pasture rotation. It is a perennial polyculture and extremely leafy. The rule of thumb in many areas is that hay needs to be replanted every seven years to keep the stand from thinning. Hay fields generally look brown after mowing because the grass clumps have a fair amount of dirt in between them due to the crowns imploding (the extremities die off and the plant shrinks toward the middle). This indicates succession going backwards due to insufficient animal impact.

At high density, the animals chip away at the grass clumps, opening up the crowns and stimulating additional tillering. As the clump edges break off, buds sprout and the clumps spread, filling in the dirt areas and tightening up the sward. Our hay fields are green like a golf course after mowing because of the sward density. On most farms it is common to see dust flying during haymaking, but again this is symptomatic of imploding grass clumps.

The best thing that can be done to a hayfield is to graze it at high densities a couple of times a year. That will kick succession in, thicken the stand, and rejuvenate it. We can do the same thing with controlled grazing that commonly requires plowing and planting. This thicker sward produces a hay that is softer: more blades relative to stems.

Generally we graze all the fields early in the spring and then begin dropping them as the grass growth speeds up and we can't keep up with it. That means all our hay has been grazed at least once and often twice. By delaying the maturity a couple of weeks, we can make high quality hay later in the season when the weather is hotter and drier — also after the meadowlark peeps are big enough to fly. See how beautifully it all fits together?

Using the cattle as a tool to delay physiological maturity, we can harvest May type hay but do it when the weather is more conducive to making good hay.

As fertility comes up, senescence (the aging process) is delayed. This also works in your favor to delay maturity later into the summer.

If hay is made well and kept dry, it will last indefinitely. I had a friend who purchased a farm and needed to clean out some 30-year-old bales of hay in a barn that needed repairs. I took some of the old hay to use as bedding. Not much of it went for bedding. The cows ate it like ice cream. It still had good color, good smell, and lots of leaves.

That reminds me of Vaughan Jones, the New Zealand forage guru, who said that well-made hay is every bit as good as silage. Certainly there are good applications for grass silage, and maybe we'll try some of that. But well-made, high quality hay is hard to beat. It doesn't happen automatically, for sure, but the difference between mediocre hay and well-made hay is enough to warrant the extra care.

Since it can keep indefinitely, good hay is a wonderful safety valve. We try to keep a year's worth of hay on hand. That sure makes the pressure of a drought easy to bear. It's like a huge savings account. If other people begin liquidating and the price of cattle slumps, we can pick up a few and take advantage of the swing. It's a good barometer to see in a tangible way how your productivity is doing. If your herd is increasing and so is the stockpile of bales in the barn, you know you're making progress. If it's going the other way, then you know you're running ahead of your fertility. It's a wonderful gauge of carrying capacity that encourages you to think long-term.

Hay is certainly not an end in itself. As I mentioned earlier, a lot of folks like to brag about how much hay they make; I like to brag about how much hay we ***didn't*** make. But a barn full of well-made hay can sure eliminate a lot of sleepless nights. It's another tool to add flexibility to the enterprise. And if you're going to make it, you might as well make it right.

Chapter 35

Winter Hay Feeding: The Nutrients

In non-brittle environments with more than 27 inches of annual precipitation and a winter forage dormancy period of three months, manure management, hay feeding, soil fertility and soil pugging require as much attention as the summer pasture rotation schedule.

Perhaps nothing brought this into sharper focus for me than what the media called the 1993 "epic" snowstorm, which pummeled the eastern United States. Here in the Shenandoah Valley of Virginia, we received nearly 20 inches of snow, winds nearly 60 miles per hour, and temperatures below zero. That was a triple whammy that left many livestock producers economically crippled and emotionally drained. Most were in the middle of "spring" calving too. While most folks here suffered dead baby calves, hard-to-start engines to deliver hay, and even inability to deliver hay in 12-foot snowdrifts, we enjoyed the warm fire, half an hour of chores each day, comfortable cows, and generally a winter vacation.

All we had to do was walk out to the hay and livestock shed, throw down some hay, and chores were done. We were not calving and didn't have to start a single engine or drive anywhere to deliver hay. The cattle stayed warm and dry on the deep bedding in the hay shed. The whole experience brought into sharper focus the beauty, the life-style aspects, of this approach to shed feeding hay on deep bedding.

Before I continue with this topic, I need to address the brittle, or arid, environment. Shed feeding may not be applicable, because pugging is not as big a problem. Also, since the forage palatability window is wider, routine range grazing is more common. By the same token, I do not think these ideas should be dismissed out of hand.

Plenty of western ranchers pen their livestock up close during the dormant season and their manure suffers the same loss as it does in non-brittle environments.

Since soil carbon, related to organic matter, is the nutrient sponge, as organic matter increases nutrient losses decrease. OM is also the key to soil structure, and therefore reduces pugging damage. As OM increases, and the sward thickens, it becomes strong enough, and resilient enough, to take wet dormant season impaction that it would not otherwise withstand. Many folks citing the buffalo and elk herds of the great plains wintering in tall grass prairie disagree with my shed wintering/nutrient loss ideas vehemently.

My response is that when we reinstitute the OM levels of tall grass prairies and the mulch tall grass prairies normally had in the winter, then perhaps these ideas will be archaic. But our farm had been plowed to row crops for nearly 200 years before we came in 1961. That is nearly two centuries of mining out the topsoil. Sir Albert Howard, godfather of composting, articulated agriculture well in his classic, *An Agricultural Testament*: "It is the temptation of every civilization to turn into cash what it took nature centuries to build." He was speaking of organic matter.

Certainly a noble goal is to be such excellent stewards of the land that we would restore the species and the soil structure that existed here before the plow and row crops. This is not to suggest that tillage cannot be practiced in a regenerative way, but in our nation's history, at least, it has generally not been. Some wonderful practitioners are growing row crops, but most are still mining the organic matter and nutrients out of the soil.

In areas where the soil freezes solid for long periods of time, nutrients are not as mobile as in areas where freezing and thawing occurs routinely throughout the winter. Total rainfall during the winter has a direct bearing on nutrient mobility.

The starting point is to establish the value of one cow-day of manure. I have never seen such a figure in any book or agriculture economics analysis, and yet it is foundational to how we view manure. Is it waste or resource?

We know that each 1,000 pounds of liveweight — roughly one cow equivalent — drops 50-60 pounds of manure and urine per day. Of that material, roughly .35 pound is N, .23 pound is P and .28 pound is K. Assuming 1995 values for chemical NPK at 24 cents per pound of N, 22 cents per pound of P and 14 cents for K, we come up with a total daily value of 8 cents worth of N, 5 cents P and 4 cents K. In many areas, current prices are higher than these. I've chosen conservative, rock-bottom prices in order to be more than fair in these calculations. In just NPK value, then, a cow-day's manure is worth 17 cents or more. Add the organic matter and trace minerals, and the value doubles. I've had fertilizer dealers tell me that really the main value of manure is organic matter. But for the sake of discussion, I'll focus on NPK.

These nutrients are highly soluble. That means if they contact moisture, they quickly diffuse throughout. On the other hand, if they get dry, they tend to vaporize — especially the nitrogen — and we can smell the nutrients escaping. If manure is properly managed, it will never smell. When you smell manure, just visualize dollar signs going off into the air, because that is exactly what is happening.

The point here is that what pops out of the cow at 30-40 cents per day is about as stable as ice in July. I think it behooves us to stop a moment and do some mental calculating as to the potential value we get from our cows each day.

If we just use the NPK figure, forgetting the biological, trace mineral and organic matter values, our 10 cows are giving us $1.70 per day in fertilizer; 100 cows $17 and 500 $85.

That adds up fast, doesn't it? Now let's take it one step further, to an annual value per cow: $62.05. That is ***just*** the NPK value of that roughly 20,000 pounds of material that each 1,000 pound equivalent is giving us every year.

Problem: these dollars are as fleeting as smoke from a campfire. Sometimes we lose very little, and other times we lose practically all. If we are running a good controlled grazing program with a thick pasture sward and vigorously growing grass with an active decomposition cycle going on we lose very little of the nutrients our cows give us. Soil microbes jump on the nutrients, transferring them into usable chelated elements that microorganisms can transfer into root hairs for plant metabolism. The more active the soil decay cycle, the quicker the incorporation, the faster the assimilation of the nutrients, and the fewer the losses.

But now let's go to the other extreme of the year. Rather than talking about May and June, let's look at January and February. The dung that drops in January typically will lose well over 50 percent of its nutrients by spring.

Remember, capturing these nutrients occurs either through quick assimilation into plant tissue or by being stabilized some other way. Some researchers suggest it is common for 100 percent of the NPK to be gone by spring, leaving only the organic matter as a benefit. Years ago, we realized the significance of these losses when we spread fresh manure in January and right next to the area spread fresh manure in April. The difference was astounding. Essentially, the winter-spread manure gave no benefit.

At that point, we vowed to keep every pound of fresh manure possible off the fields in January, February and March, resolving to stabilize them some other way in order to hold the value.

It is outrageous that so many liquid manure pits get pumped and spread on frozen ground in the winter because that is when farmers have the time and can get on the ground. It is especially bad when it is spread on frozen ground on top of a snow. Where are all those nutrients going to go when the thaw comes? Such losses are especially detrimental because the farmer loses the nutrient value and the nutrients go where they are liabilities rather than assets. If they simply moved from one place to another, but could still be used, that would not be as bad. But to add insult to injury, they pollute streams and eventually bays and estuaries. Everything loses.

The primary culprit is leaching. Because the nutrients are extremely unstable and the soil microbes cannot metabolize them during the dormant season, winter rains and snow washing across those manure pats leach away all the nutrients. The more severe the winters in terms of freezing and precipitation, the greater the losses. In other words, this is a more acute problem in Maine than in South Carolina. In some northern states it is illegal to spread manure during the winter because monitoring nutrients leaching into lakes and streams has clearly substantiated the tremendous instability and losses of these farm nutrients. Spring, summer and fall do not yield similar figures; winter is the key to reducing farm nutrient losses.

We find minimal losses on nutrients applied (including droppings during grazing of stockpiled fescue) up until mid-December because the soil microbial activity hasn't completely gone to sleep. But once those hard freezes begin it's a different story. The soil is the earth's stomach. We should not attempt to feed it while it is asleep any more than we would get up at midnight and eat a Thanksgiving dinner.

Old-timers around here tell me that one of their routine chores during the winter was to go around the barnyard and pick up cowpies with a fork. They would put them in a wheelbarrow and then take them into the barn to add to the animals' manure pack. That would protect the nutrients until they cleaned out the barn in the spring.

The 100 days from mid-December until mid-March are the greatest challenge for holding onto nutrients. Pasture droppings during that time are virtually wasted. To add insult to injury, that is when we do the most pugging damage. Fall pugging damage recuperates fairly well during the January-February freezing and thawing period because of the way the ground heaves repeatedly. But pugging during the freezing-thawing period (our freeze line is a little more than two feet — more temperate areas do not have as acute a problem with pugging), or between freezing and grass green-up, can ruin an area for a couple of years. Many folks use a "sacrifice" area for this time, but I don't want to sacrifice any area and I certainly don't want to sacrifice any nutrients.

Tragically, most of what has been written about this period in the agriculture press concerns calving, nutritional requirements of the cattle, supplementary feed and nutrition. Certainly these are important, but by the same token a 17 cents per cow-day loss in nutrients deserves some attention.

If you have 100 cow equivalents out in a muddy lot imagine just taking $17 a day out of your wallet and burning it. The tragedy is that while you are losing the nutrients, they are showing up in both surface and ground water resources as nonpoint pollutants, or filling the air with foul smelling ammonia.

Let me illustrate the magnitude of the problem this way. A normal productive pasture in our area will metabolize up to 150 pounds of nitrogen per year. The Soil Conservation Service only allows 140 when formulating nutrient management plans for factory poultry farms.

A dairy or beef operator has a drylot of three acres holding 90 cow equivalents. Those 90 cow equivalents are dropping 30 pounds of nitrogen per day (1/3 pound per animal). If it were evenly distributed over the three-acre lot, which it is not, of course, that would be roughly 10 pounds of nitrogen per acre per day. In just 15 days, enough nitrogen would be added to the soil to fully saturate it for the entire year.

Once an area is saturated, what happens to the nutrients applied the next day? They vaporize in the air or leach into the water. Period. The soil is the stomach; it is full; it can't hold any more. Of course, normally a loafing lot like this will not even be healthy pasture; it will be devoid of grass, so it can't even uptake a normal dose of nitrogen.

I hope this example helps us all to appreciate how much nitrogen specifically, and nutrients in general, are being wasted each year in this country. In his *Complete Book of Composting,* J. I. Rodale made the classic observation that there is no reason to buy one ounce of chemical fertilizer in this country. Enough nutrients exist on farms and at the back end of tanneries, food processing establishments and municipal lawns and garbage bins to more than supply ***all*** the soil nutrient requirements of the entire country. The problem is that most are wasted, inefficiently utilized or buried in landfills.

As early as the 1920s Sir Albert Howard envisioned what he called "garden cities," where 25 homes would be built in a diamond shape so each could have solar access. Each 25 families would employ a master gardener and composter who would compost all their human and kitchen waste, growing all their vegetables and fruits inside the diamond. He thought within a few short years the world would see the folly of water-based manure management systems and employ composting.

Here we are, nearly a century later, still denying building permits for composting techniques and requiring water-based sewage systems. We're trucking vegetables from the salinized valleys of California and sending eggs halfway across the country. It's amazing how creatively we can mess up things.

These losses also explain why so many farmers, as noted before, receive five tons of manure per acre per year as a total-farm average, but spend thousands of dollars on fertilizer. How tragic.

Manure pits are probably the worst way to handle manure. The negatives of slurry systems are significant:

(1) Nutrients remain highly soluble. Nutrient instability increases opportunity for leaching and vaporization.

(2) Water is expensive to haul. One of the chief problems faced by farmers is soil compaction brought on by traversing the ground with heavy equipment. In order to justify running to the field with a lot of water and a few nutrients, high-volume, heavy tanks are the only efficient way to go.

(3) Ammoniacal nitrogen burns out organic matter in the soil. Fields routinely covered with slurry show marked organic matter declines unless very aggressive cover crops and high carbon resi-

dues are tilled back in.

(4) Storage often is inadequate, causing spills. Especially in high precipitation winters, lagoons fill up but emptying is out of the question due to weather. Heavy precipitation exacerbates the situation. Spills often end up running down a creek.

(5) Spreading is a logistical nightmare. The high precipitation months are in the winter, when spreading is an environmental disaster. At least one government official who implemented the lagoon program, wishing to remain anonymous, told me more nutrients were washing away with pits than when manure was being spread fresh every day. Much of the slurry is spread on snow or frozen ground, when the dormant soil cannot metabolize the nutrients. Farmers are busy with other things in the summer, when it would be best to apply. As a result, much, if not most, of the material is spread at inappropriate times of the year.

(6) Bad public relations. Because the material is so volatile due to ammoniacal nitrogen and high solubility, warm weather spreading generally causes incredible odor problems. This upsets the neighbors and further polarizes farmers from the community. If manure is properly managed, it will never smell. The loss in nitrogen, of course, is a direct nutrient and economic loss to the farmer—on a warm, sunny day losses can be 30 percent of N in 24 hours.

(7) Expensive upkeep. Pits are depreciated just like buildings, requiring maintenance and eventual replacement. They are capital intensive systems, often exceeding $25,000 in cost, even for small dairies. The equipment required to stir, pump and haul is also expensive and maintenance-demanding because of the slurry's corrosive qualities.

(8) Farm safety. How many farmers and children have died accidentally falling into a manure lagoon? These systems require "heavy metal" to function, and are prone to safety problems.

If we would put as much attention on managing the manure as we do on developing better corn, or even managing our pastures, we would make more money and be far more environmentally enhancing. It's time for agriculture to understand the challenge and act more responsibly.

One sidelight is that these nutrients are worth far more than face value, because the dollar you save is not taxed. If you have to earn the dollar to buy back the loss (i.e. lose 17 cents in nutrients and spend 17 cents buying them back from the fertilizer dealer) you pay taxes on those earnings. The IRS has not yet devised a way to tax our cow paddies. Every time I see a cow hunch up and do her business, I say: "Thanks for the dollar, Bossy. I appreciate that investment and the vote of confidence."

The nutrients are valuable and they need to be tied down, for the benefit of the farmer and the folks downstream.

Chapter 36

The Hay Shed: Nutrient Capture and Cow Comfort

How do we tie down the winter nutrients and virtually eliminate these losses?

Carbon.

Very simple. When the highly unstable nutrients contact carbon atoms, the two parts bond molecularly in a chemical sponge. This stabilizes the highly soluble nutrients so they can be retained until soil microbes awaken enough to assimilate the nutrients quickly.

Carbon, or carbonaceous material, comes in many forms: sawdust, wood chips, paper, straw, leaves, junk hay. Generally known as biomass, high-carbon material is available from many different sources. Every area of the country has available material. Here is a list of possibilities:

- Leaves from the municipal leaf dump.
- Sawdust from a sawmill or wood processing facility.
- Wood shavings from cabinet or furniture makers.
- Wood chips from municipal dumps or from rights-of-way chipper crews. We pay a crew $10 a load to dump here at the farm and they bring about 25 truckloads a year.
- Christmas tree chips. When our town chips its Christmas trees, we drive our 15 cubic yard dump truck in and they blow them right in. In a day we can haul home four or five loads.
- Cotton ginnings.

Photo 36-1. *Carbon drives the fertility engine. Stockpiling biomass for bedding material is done whenever we find a "good deal."*

- Peanut hulls.
- Wood chips you generate yourself. A PTO-powered hydraulic-feed chipper can be purchased for $7-9,000 that will handle material eight inches in diameter. In short order you can turn branches into piles of useful chips, or utilize a woody area that needs to be thinned or upgraded.
- Horse stable cleanings. We clean a big stable and haul the material home for hay shed bedding. Horse people are so meticulous that this material is hardly soiled and can be reused.
- Bark peelings from sawmills or other wood processors.
- Straw. Ideally, small grain straw should be rained on a couple of times before baling to knock off that slick exterior. Dull straw is softer and more absorbent.
- Junk hay. Old round bales or weathered or rained-on hay can often be had for 50 cents a bale plus hauling.
- Vegetable processing wastes. Tomato vines, grape stems,

apple leaves, etc. are often available if you live near a processor of these foodstuffs.

- Shredded cardboard, newspaper or office paper.
- Wood straw from chicken shipping boxes or packaging material.
- Corn fodder.
- Corn cobs.
- Grain husks from grain processing facilities or seed companies.

Enough carbon must be put in contact with the manure to chemically suspend the nutrients. The critical formula here is the carbon:nitrogen ratio, or C:N for short. As long as we keep the C:N ratio around 25 or 30:1 (i.e. 30 parts carbon to 1 part nitrogen) we can keep nutrients captive. As it drops below 25:1, the material wants to release nitrogen, which is great if we are spreading it as a fertilizer during the growing season to grow a crop of hay or corn. On the other hand, if we spread material that is 40:1, say, we will have a temporary nitrogen depletion in the soil until further decay brings the material down to 25:1 and it begins giving up nitrogen. This is a simplistic picture of what happens, but it explains why we can affect the available nitrogen in the soil, and hence its crop-growing abilities, by the type of material we apply.

Just so you can visualize what this C:N ratio is, let me give you some figures:

Poultry manure	7:1
Cattle manure	18:1
Leaves	50:1
Offal (slaughter waste)	2:1
Straw	100:1
Sawdust	500:1
Wood chips	250:1
Grass hay	80:1
Legume hay	20:1

To maintain a 30:1 ratio requires different volumes of carbon depending on the material you use. For example, to tie down one cow-day's (50-60 pounds) worth of manure and urine would take far more leaves than sawdust. This is easy to visualize. Working out the relationships is simple math. If you have one pound of cow manure at 18:1, addition of one pound of sawdust at 500:1 would yield a C:N ratio of 518:2, or 259:1. Although one pound of dry sawdust may not sound like much, compared to one pound of cow manure (roughly one pint) it would be a lot — a gallon or more.

If you have 10 pounds of cow manure at 180:10 and you add one pound of sawdust at 500:1, you have 680:11 or about 60:1. This is why dry sawdust goes so far in a livestock bedding situation, compared to something like junk hay. It takes far more weight to get the same nitrogen-retention out of junk hay than it does out of sawdust. Of course, a pound of dry sawdust could easily be a cubic foot of volume, whereas a pound of hay would only be a quarter of that volume.

Depending on what material you have, then, you can figure out what you need to tie down the manure. The rule of thumb is that if you smell ammonia, add carbon. Just like grazing management, experience is the best teacher. You want to stay right on the edge: just enough to tie down the nutrients, but not too much. Obviously the challenge is to keep enough carbon on hand. Putting in too much is not really a problem.

Besides carbon, soft rock phosphate and potash (wood ashes) added to the material can also help tie down nutrients. In fact, some research shows that minerals (including calcium) applied with compost go up to four times as far as when applied separately. In other words, 500 pounds of ground limestone applied as part of compost — especially if it was part of the biological activity of the composting process — is equivalent to a ton applied by itself. Something to think about.

Handling this carbon involves labor and investment, but the carbon itself contains many nutrients. For example, just the NPK value of leaves is $5.48 per ton (assuming a pound of N at 24 cents,

P at 22 cents and K at 14 cents). Wood chips are worth nearly $5 per ton just in fertilizer value. Grass hay is worth almost $10 per ton just in NPK value. And again, the real value of carbon has nothing to do with NPK, but rather with feeding the decay cycle in the soil, which is the secret of soil fertility. About one ton (5 cubic yards) of dry chips will tie down the nutrients from 200 cow-days' worth of manure and urine.

The biomass must be dry when the manure and urine touch it, and then the combined bedding must be protected from rain. Dry sawdust, leaves and wood chips will absorb about 200 percent of their weight in moisture. Cattle droppings are 60-80 percent moisture. If the bedding goes in wet, it cannot absorb this moisture.

The celluloid spaces of the bedding material can be displaced by water or urine; the material will only absorb so much. If the water-retentive capacity is already filled before the bedding is put down beneath the cattle, the urine, which contains the lion's share of the nitrogen, will just leach through it and into the ground or off the cement.

Photo 36-2. *The hay feeding shed. Note hay storage in free-span middle (45 feet square) and feeding awnings off each side. Very open but protects winter dung and urine from leaching.*

Most biomass materials begin wet, though, either with rain or sap. By piling the material under roof, it will heat and the fledgling decomposition process will dry it out. It almost becomes fuzzy and soft with the molds. Once you see the steam coming off the top of the pile, it takes about two weeks for full dry-down.

The cattle loafing area must be roofed also, to protect the bedding from the rain. Typically 10-20 inches of precipitation will fall during this 100-day period. Imagine how much dry carbon would be required to soak up five gallons of urine per cow per day, plus that much water. It would be virtually impossible. If you succeeded, the C:N ratio would probably be 50:1 instead of 30:1. We don't want to handle any more biomass than necessary.

We built a shed out of our own locust poles, even using locust poles for rafters, and scavenged roofing stringers from a neighbor's barn demolition project. Total cost for materials (bolts, roofing, nails) was 46 cents a square foot. On one side is a V-slotted feeder gate that can move up and down to accommodate the bedding pack buildup. The 20 foot high hay shed is flanked by this feeder gate and the lounging awnings. The distance from feeder gate to the lounging area back wall needs to be 16 feet or more in order to give enough room behind the eating cows for timid animals to come in and find their spot. If only a few feet separate the rear end of feeding boss cows and the back wall, smaller cows will not pass on down the line to find their place in the gate

The feeder gate is four feet high and the 'V' slots are on roughly two foot centers. Dry cows are actually a little wider than that, so we try not to put more than 21 cows in a gate with 23 slots. The gate can raise and lower four feet to accommodate bedding buildup. The deeper the bedding, the better it works. Also, we do not want to clean out during the winter because an ambient temperature of 50 degrees Fahrenheit is necessary to begin compost. We need tremendous storage capacity. By spring sometimes the cows are rubbing their backs on the awning roof, but at least everything is protected.

We add fresh bedding as needed. Although the cows are free to go outside into a small loafing area, they spend 80 percent of their

Photo 36-3. *Cattle loafing area under awnings of hay feeding shed. Awnings must be at least 16 feet wide. Deep, clean bedding keeps cows clean, dry and warm. Shed protects cows, manure nutrients and hay.*

Photo 36-4. *Cattle eating in the V-slotted feeder gate. Feeding is simple: no diesel engines to start and no delivery. Just drop hay in front of gate and let cows do the work.*

Photo 36-5. *Note bow truss on feeder gate for added strength and rigidity. Cows can push hard when they lick up the final morsels of hay. Gate protects hay from soiling, thereby protecting cattle from parasite ingestion.*

Photo 36-6. *Side view of feeding awning. Note buildup of bedding. Feeder gate can raise and lower four feet to accommodate bedding height. Chips for bedding stored adjacent, at extreme right. Whole system significantly reduces hay consumption while increasing cattle performance.*

Photo 36-7. *Adding corn for pigaerators to bedding material in manure spreader.*

time in the awning because it is warm, clean and dry. In a wet period, we may have to add bedding every day, but during a dry period we can get by with once every three days. Cows obviously drop more dung than calves and need more bedding.

Lounging space should be 30 square feet per cow. At our cost (excluding labor) that comes to $15 per cow. That is the cost of shedding her in the 100-day nutrient-loss, bad weather period. If that gains us 17 cents worth of nutrients per cow-day during that 100-day period, we've paid for the shed in one winter in nutrient saving alone.

If the simple pole facility costs $3 per square foot, the payback is about 5 years; $4 gives a 7-year payback and at $5 a 9-year payback.

How many capital investments can you make on the farm to provide that sort of payback? When round bales came to Virginia, and the ag economists realized 50 percent of it was rotting before it ever got to the cow, they figured the value of hay and price of storage. The bottom line was you could justify building cheap pole storage in hay savings alone. That is similar to what I am attempting to do here.

But let me answer the critics' inevitable arguments.

What about interest for the money to build a shed? That is handled by the fact that you won't have to earn and be taxed on the money to buy (replace) those lost nutrients. Further, does anyone think petroleum-based fertilizer will get any cheaper?

What about the labor? With this system, one person can feed 150 cows in half an hour by hand without ever starting an engine. No ruts, no driving out in disagreeable weather. No diesel fuel for hay delivery.

What about the cost of handling and acquiring the biomass? Fertilizer isn't cheap. The biomass has tremendous intrinsic value to feed the decay cycle, and what else are you doing during this time of the year, anyway? If you only spend 30 minutes feeding the cows, you have plenty of time to do other things. Front end loaders can put quite a bit of material down in a hurry. A slick way is to put the bedding in a PTO manure spreader and drive slowly through the lounging area, spinning off the sawdust or whatever. You can chip your own material into a silage wagon and unload it at the storage site.

Photo 36-8. *Applying bedding is easy with PTO manure spreader. Just drive slowly through the loafing area.*

Biomass can often be had for the hauling. Look around. Carbon acquisition is just part of our routine like baling hay. We spend about five days a year hauling in carbon.

And now the big one: "You're only dealing with a hundred cows. We have five hundred. We couldn't afford to build a shed that big." This argument is the typical: it works for you little guys but it won't work for us biggies. Notice that all my figures had nothing to do with large or small: they were all given on a ***per cow*** basis.

In other words, the bigger the operation, the more it will cost to get a handle on nutrients, but the more gain can be realized. My total daily nutrient value is only $17, for example, while somebody with five times as many cattle receives a daily nutrient value of $85. The bigger the building, the cheaper it is per square foot, so actually per-animal costs go down with the larger herd.

The numbers are all relative and the bottom line is size neutral. The nutrients are being dropped per cow, regardless of numbers. I could say I can't afford to do this because I'm only taking in thousands of dollars in gross sales while a big operator may take in hundreds of thousands. Looks to me like the biggie can afford it easier.

Of course, it is as silly for me to think the biggie can afford it easier as it is for him to think I can do it easier. The numbers, nutrients, and principles involved have absolutely nothing to do with size. The benefits — as well as the costs — are size neutral.

Finally, I failed to articulate, on purpose, a couple of tremendous benefits that I haven't put a dollar value on, but which certainly do have an economic value.

First, the bedding stays warm as it ferments (anaerobic) and this provides the cows with a heated bed. That means the cows stay warmer, requiring fewer calories during cold weather. Not only do they stay more comfortable, but they also require 5-10 percent less hay to maintain condition. You will feed less hay.

The feeder gate remains sanitary from their droppings, unlike hay fed in the pasture where the cattle walk on it. This reduces parasite infestation. Furthermore, there are no pugging areas in the pasture to overseed and reestablish.

Less hay is wasted because the cattle can't pull the hay out of the feeder gate. They must put their heads in the 'V' at the top and eat out of the deep box on the other side. Hay wastage is virtually zero.

The deep bedding produces molds and fungi that in turn produce natural antibiotics to maintain health. The deep bedding will reduce lice and other pathogenic organisms.

Gone is the entire pugging problem, muddy cows and unsanitary conditions. Since the cows' hides stay dry, because of a clean lounge area, their insulating ability functions optimally. When the hide of a cow gets wet, she loses insulation and becomes more susceptible to sickness. She also must consume more feed to maintain body temperature. Again, the hay shed and deep litter reduce hay requirements.

Neighbors are happier. Nothing raises the ire of animal rights activists and environmentalists more than uncomfortable-looking cattle and muddy feedlots. The cattle gravitate to the shed like a magnet, which protects the fields or lot from mud and impaction. It has every aesthetic quality we could want, allowing us to be friends with city folks.

Although I do not have a quarrel with the many graziers who have opted for no-building wintering, these concepts and economic realities merit examination. And while it certainly is not applicable to everyone, it may be more applicable than many of us realize. Wintering the cows "down in the woods" may be easy, but when we consider hay savings and nutrient load, is it also economical? We cover roughly 25 percent of all the pasture each year with the compost generated from this 100-day winter period, and I can vouch for the efficacy of compost over any other soil amendment.

If we are going to make our farms profitable and environmentally sound, we must look at manure as a resource, not as a waste. Our management system should be designed around multiple use, getting the most benefits from the actions involved. Slurry systems are hard on the cows because they require concrete; hauling fresh every day smells up the neighborhood and applies many nutrients at the wrong time. The list articulating the negatives of water-based or non-compost manure management systems could go on and on. Even the no-management (winter them down in the woods) is not a solution over the long haul.

When we list the positives of protected hay shed feeding, from a labor, environmental, economic and animal comfort standpoint, it is a multiplicative value, to be sure, and well worth the effort of a simple building and some biomass management.

Chapter 37

Pigaerators and Compost

The carbon:nitrogen ratio should be about 30:1 for good composting, and if we have done our preliminary work correctly in the feeding shed, we will be about right.

The rule of thumb is about a third nitrogenous material and two-thirds carbonaceous material. The reason biomass is so important is that we cannot compost manure without it. If we scrape a concrete floor every day, those manure scrapings will not compost because they are too high in nitrogen. Composting will stabilize the nutrients to minimize losses once we take the bedding pack to the field.

Composting reduces solubility of NPK by 30-40 percent. This stabilizing effect keeps the nutrients on your farm rather than letting them wash off and go down the creek or into aquifers. The composting heat kills pathogens which otherwise reinfest the cattle. It also kills weed seeds.

Composting eliminates repugnance that accompanies applications of raw manure and cattle droppings — including the bedding pack. You can spread compost one day and graze over it the next. The cattle are not bothered by it in the least.

Composting is a skill, a combination of art and science, just like graziering. The coarse and more carbonaceous the biomass material you use for bedding, the longer the breakdown. It's also easy to have too much carbon. We use junk hay or straw every fourth time we bed down to make better compost. The tender material reduces the spiking heat of the pile, mellowing out the temperature and making a better finished product.

Do not become overly concerned about perfect compost. The goal is to stabilize the material. If it's a rough compost (not fully decomposed) the soil organisms will finish it. Let time be your friend rather than your enemy. The material should at least smell like forest soil or moldy leaves, not like manure.

The heat should not exceed 160 degrees Fahrenheit. Too much heat can burn out valuable nutrients. If the pile looks white inside, that indicates fire-fanging. This occurs when the pile gets too hot and actually cooks out the nutrients. The cure is to reduce the coarseness of the carbon and/or add water. If the pile won't heat the material is probably too wet or too low in carbon. Add more carbon and remix.

Opportunities abound for composting. In essence, composting diversifies the farm into fertilizer production. We also compost our beef offal, bringing the guts back from the slaughterhouse. This is wonderful fertilizer, and illustrates how many sources for compost materials exist.

It just doesn't make sense that to maintain fertility we should have to import material from 100 miles away. I don't believe God designed things that way. Ed Faulkner, in his classic, *Soil Development*, showed how constant addition of organic matter to a totally sterile soil can quickly produce a completely fertile, productive soil without the addition of anything from outside. Native forest soils and prairie soils certainly did not depend on any bagged, imported material. The soil complex, if it is enjoying an active decay cycle, can solve its nutrient deficiencies. A cow can eat nutrient-poor forage but excrete nutrient-laden goodies.

Biodynamic practitioners talk about transmutation. Science scoffs at the notion. But truly soil microorganisms can work magic if we feed them enough of the right things.

Using pigs to aerate and mix deep bedding revolutionizes the economy and efficiency of large-scale composting. A net profit of $300 per hog can be realized if the hogs substitute for labor and equipment.

Because the cattle tramp out the oxygen in their deep bedding, it is anaerobic and therefore toxic to aerobic soil life which lives in the top 3-6 inches of soil. To be truly useful as fertilizer, the bedding pack should undergo aerobic composting before being spread on the field. Typically, this is done with a PTO-powered manure spreader inching forward building a windrow, mixing, and injecting oxygen all at the same time. We used this obsolete method at one time too.

"Pigaerators" can accomplish the same task and you won't have to start up a tractor at all. During the hay-feeding period, we put down corn and oats before adding fresh bedding. At the end of the feeding period, the nearly 4-foot deep bedding pack is full of fermented grain. We use small grain early and corn late because the grain inventory is first in, last out. Corn in the bedding for more than 60 days or so will ferment away to nothing but skins. It takes roughly 30 pounds of grain per cubic yard of material to feed the pigs.

When the cows come out in the spring, generally around the first of April, we purchase hogs and turn them loose in the bedding pack. They are more entertaining than the Three Stooges. We simply connect a hog nipple waterer to a pole and gravity feed water from a hanging bucket.

The pigs start with the easy grain up near the top, and root it all at that level. Then they have to go deeper. They prefer the fermented grain to fresh. Once they get down close to a foot, the material they've been through is heating well. Of course, they add their own manure and urine to the pile.

The fermented kernels of grain should be intact, but the insides can be fermented into white juice. Two pigs will turn 75 cubic yards of material in 8 weeks on 500 pounds of grain per hog. Figure on losing about 100 pounds of grain per hog — they won't find every kernel. They move through the material like an eggbeater. Using their snouts, they sling the overburden around behind them as they go through the strata of fermented grain, kind of like a surface mining operation.

No more than two hogs should be put together at a time because as they go deeper, beyond two feet, only one pig will dig and the others will stand around the hole eating the grain the digging pig throws out. He eventually tires and gives up because he is not getting enough energy to feed his engine. Apparently pigs like welfare handouts too. We buy large pigs — at least 150 pounds — to start, and take them on up past 300 pounds.

The deeper they go the stronger they need to be. They naturally grow as they get deeper and deeper. The multiple stirring makes a far superior compost to windrowing. The reason farmers shy away from large-scale composting is because of the materials handling required. But when animals do all the work, that becomes a non-issue.

In addition, pigs do not require a minimum wage, do not break down or require mechanics and parts; they don't even require Social Security! In fact, when they are finished we enjoy eating them — what a retirement plan! Edible farm equipment. Instead of rusting in the field when it's served its time, we serve it up for dinner.

This whole paradigm provides wonderful opportunities for humor. I was in the farm store recently purchasing some hog nipples. The clerk, knowing my reputation for frugality and penchant for profit, raised his eyebrows and queried: "What are you doing buying hogs? You can't make any money on hogs."

At the time, the hog market was so poor that a farmer couldn't steal the pigs and steal the corn and make any money. But I shot back: "Oh, but I'm not buying hogs. I'm buying 30-cent per pound tractors and farm equipment."

He paused for a moment, then thoroughly confused, asked: "Do you ride these hogs?"

I couldn't keep leading him on like that, so I had to break down and explain what we were doing. All his premonitions that I was a lunatic were confirmed, but it sure made a good story at our family supper table that evening.

When we let animals do the work, it completely changes the economics of the farm. The cost of the pigs and the grain they need directly offsets the machinery cost and labor of double handling the material to build the windrow compost pile. When they are done, we have pork to sell, and it's virtually all profit.

Anyone who has cleaned out a cow manure pack knows how tight it gets. But after pigs have composted it, the material is light, fluffy and can be loaded easily into the manure spreader for pasture application.

Letting animals do the work takes the labor and expense out of large-scale composting. And it's just plain fun.

Marketing the Salad Bar

Chapter 38

Relationship Marketing

In light of the insatiable human desire for empire-building, perhaps it would behoove us all to back off and examine the merits of relationship marketing.

While economics lessons generally laud practices that increase business size, business volume and annual quantitative growth, equal time should be given to the notion that numbers do not equal profits — or happiness. I was fascinated recently with an article in one of those freebie airline magazines about time and work. Twenty years ago, experts predicted a 30-hour work week, more leisure and a Jetson's life-style of button-pushing and console monitoring. But alas, people are working more hours per week at stress levels that eclipse those of just a few years ago. We are in fact working more and enjoying it less.

While not every farmer can sell products at retail prices directly to the consumer, many more can than can't. By selling only to those who will come to the farm, the producer will inherently remain small compared to empire-builders who pack, ship and advertise all over the country. While we do not have thousands of customers, our several hundred are more than enough to provide the income and life-style we desire.

Our marketing is not so much moving a commodity, but rather a natural outgrowth of building relationships with people by meeting their needs. What people need is good, clean food at an affordable price. That is a timeless need.

The first advantage of relationship marketing is consumer education. In practically every agriculture publication I receive from the conventional crowd there is some article about those stupid consumers who don't understand this thing or that thing. Consumers are considered ignoramuses, constantly trying to "do the farmer in" by supporting ludicrous social and legislative policies. But instead of opening our farms up to them, we post big **NO TRESPASSING** signs at the entrances. How can we educate consumers, how can we build a bridge to them, how can we garner their trust if we shut them out of our farms and call them names?

Consumers are just like producers. They are inquisitive people who reciprocate treatment in kind. They are sincere, honest folks who can't be held responsible for their ignorance, and who need to be pushed to learn something. Don't miss that last phrase. They need to be pushed to learn.

If we as farmers continue to market in a way that allows consumers to be uneducated about their food, are we not partially to blame for their ignorance? The fastest way to educate consumers is for farmers collectively to quit selling their wares through channels that inherently distance the consumer from their place of origin. I realize this idea has practical limitations, but it is still a valid concept and one that farmers can do their best to adopt, one at a time.

How in the world can the city consumer know what the farmer needs? He won't read agriculture publications. As the farm community continues to shrink and the consumer crowd grows, as the gap widens and mutual understanding decreases, how can the farming community possibly expect to hold its own? It will not. It cannot.

Is it too much to ask consumers to learn? I do not think it is too much to ask that a family that will routinely spend a week and $1000 or more getting in touch with Thumper and Bambi at Disneyland spend a couple of hours a year getting in touch with their food supply. Not too long ago farmers peddled their wares at roadside stands, farmers' markets, and as door-to-door vendors, engaging customers one-on-one in educational, bridge-building conversations. That has changed for the most part, especially in livestock.

And what we have now is an agriculturally illiterate consuming populace with no link — mentally, emotionally or physically, to their food. We farmers, fiercely independent and individualistic, must certainly share the blame because we have turned our products over to Madison Avenue for marketing.

We have voluntarily turned over this segment to nameless, faceless giants that truck the food across the country to nameless, faceless people. Such a system encourages separation and ignorance. It also encourages arrogance because there is none so proud as he who is ignorant of the other person's problems. This whole scenario is as devastating to the farmer as it is to the non-farmer.

This educational process is absolutely essential to keep the cities from destroying the country. Mutual appreciation, instead of mutual mistrust and fear, is truly a goal worth pursuing. For sure the consumers will not initiate the process. Farmers, one at a time, must try to regain the ground lost.

The second advantage of remaining small enough to continue relationship marketing has to do with product quality. Dad always reminded me that the biggest temptation of any growing business is to compromise product quality. Cutting a corner here and cutting a corner there add up much faster when more corners are involved. When only a few units are produced, a dime here or there doesn't add up to much. But many units, times a dime, can mount up in a hurry.

If all the large alternative agriculture producers who have gone belly-up in the last five years were tallied, it would be mind-boggling. The reasons for failures are myriad. Some were due to bad financial decisions. Many were due to disease/pest problems as the operation grew. This has been especially rampant in the animal sector of alternative agriculture. One lamb doesn't make a flock. Pets are always prettier and more productive than commercial, working animals.

Ed Martsolf, whose "A Whole New Approach" seminars deal extensively with building community, articulates a fascinating concept used by the Amish communities for making a decision: “Is this best for the community?” Most of us do not measure our decisions against the community. Yet each person needs the boundaries, the discipline, the restraint, that a community inherently brings on the actions of the individual.

When our customers’ children run between our legs, and when our patrons walk out to the field to enjoy the cows, we are forced to ponder each decision in light of what it does to ***them***. Too many farmers hide behind the veil of corporate giants and USDA approval, as if that absolves them of any responsibility for tainted food, cholesterol-laden meat that induces heart attacks, and the like. For those of us who do not have the advantage of an Amish community, the community principle is no less imperative for correct decision-making. It is a wonderful protection from improper thinking and action. If we do not have it, we must create it.

I wouldn’t want to do anything that would jeopardize the future or health of the many children that visit the farm and talk to the cows; that help their parents put packages of beef into a cooler. Such closeness requires that my question be: “Can it possibly hurt anything?” instead of the normal question: “Can you prove to me it will hurt anything?” My community demands that I err on the side of safety rather than risk. No person should be so arrogant as to assume that he is immune from the temptation to cut corners and produce something less perfect than it could or should be. Dollars are extremely tempting, and require a strong antidote. A community of patrons is a strong deterrent to the shortsighted view, a strong anchor against risky production models.

Because of our patron community, we do not flirt with things that the conventional crowd deems acceptable because they haven’t been “proven” unsafe. We want to stay as far away from risky materials and procedures as possible, not see how close we can come. As soon as the farmer is removed from the names and faces of folks eating his produce, he is removed emotionally and mentally from the impact of his decisions. This removal of responsibility awareness, whether on purpose or not, is what makes farmers, without a twinge of conscience, use hormone implants, feed poultry litter and let cows stand in the mud in the wintertime.

If the average American knew how beef — and poultry and hogs for that matter — are raised, he would be a vegetarian. That is tragic, because it need not be that way. The misunderstanding, mistrust and paranoia — on both the part of producers and consumers — is a terrible thing for our culture to have. But the kind of linkage, the kind of relationship necessary to solve these problems, does not occur from mega-business, centralized agriculture. The "get big" mentality has fostered corner-cutting and inappropriate procedures that would never have been done under the careful scrutiny — relationship — of both producer and consumer.

Nothing encourages integrity like close ties.

The third advantage is customer loyalty. It is axiomatic that customer loyalty grows as the link between producer and consumer shortens. If all the consumer knows is a label, then it's fairly easy for a brighter, snappier label to win him over. If all he knows is the advertisement, a sharper ad with a little more pizzazz can win him over. But when he knows the producer personally, suddenly the relationship element comes into play and nothing short of total disenfranchisement can pull the customer away.

The relationship builds those emotional and physical ties that are not easily broken. This loyalty cannot be measured in dollars and cents. It can't be put on a balance sheet. The cornerstone of a successful business is repeat patrons. A business cannot survive if customers don't continue to purchase. The customer turnover rate is directly related to the loyalty rate. The greater the loyalty, and hence the satisfaction, the higher the likelihood of repeat business.

We've had patrons offer us loans. "You know, I've got some money in an investment account and it's not doing very well right now. Do you need any money for anything?" We've taken advantage of this and cut several percentage points off borrowed money. We win, the patron wins: it's a win-win arrangement. We've been given money when we've had especially rough times: "Here, put this in your pocket ($100 bill). You need to stay in business for me."

Farmers who have never experienced it cannot imagine what it means to have a person, with two children in tow, look you in the eye and say: "I can't tell you how much I appreciate what you do. We count on you and just thank you for producing this wonderful food. Our family simply will not eat anything else." Anybody ever hear that from a sale barn crowd? They're usually too busy complaining about low prices, flood, drought, and stupid city people.

When the consumer is removed from the producer, and is not knowledgeable about the farming practices, a seed of doubt regarding product integrity, business ethics and the like can easily send the customer looking elsewhere. But if he has visited, seen firsthand, and participated in the farm operation, he will have the knowledge necessary to withstand some unfounded rumor.

I can't imagine why a business with an educated, loyal local clientele would want to trade that off for a nameless, faceless, capricious clientele from the four corners of the globe. Perhaps that is why many good small businesses grow up to be bad big businesses. And in the transition, many fail.

Distance, both in time and knowledge, builds an inherent gap between consumer and producer. Producing for a local or regional market, while not an option for everyone, is certainly an option for many, if not most. That bridge can and should be built. It's worth it.

The question invariably comes up after a discussion like this: "What about New York City?"

My response: "What about New York City?"

Then they begin to press: "Well, they all can't come out into the country and know their beef producer personally. That's impossible."

My response: "Why should we have a New York City?"

The point is that New York City is a flawed model, just like feedlot beef and confinement factory chicken. For the clean food community to think that it is our responsibility to ensure that an inappropriate model be sustained with appropriate food is analogous to maintaining feedlots with clean or organic grain. If a model is flawed, it should not be propped up with crutches. Let it disintegrate, and even bend to the needs of the land. This is consistent with Wendell Berry's eloquent defense of the notion that "there are no global problems. Just local ones."

Furthermore, if everyone who could sell in their communities would do so, it would open up new and heretofore unknown opportunities at other levels. The old adage: "If everyone who could, would," applies to this argument. If everyone who could market to their community would do so — and by the same token if everyone who could purchase from a local producer would do so — the cultural paradigm shift would be as enormous as a California earthquake. New markets, new shopping patterns, new thought processes, would enter the culture so that opportunities we cannot conceive of today would present themselves.

After a talk I did during a farm conference in Phoenix recently a fellow asked: "Okay, this all sounds good for Virginia, but could you apply it to Arizona? "

I responded: "Sir, Virginia doesn't think it applies."

Virginians certainly are not endorsing these concepts. The tragedy is that in areas where this whole salad bar beef concept, from production to marketing, could be so easily implemented the agricultural community is still practicing Neanderthal techniques and the consumers are still as ignorant as a fence post. True, some people live more than a hundred miles from town, but a lot more don't than do.

It's amazing how many folks hear these concepts and can quickly imagine how easy it is for me to do it, or how easy it would be for Uncle Bill, or that acquaintance up "yonder," but never make the connection that it can work for them. A lady I was talking with about this recently was enjoying the discussion but suddenly responded: "You know what? I could do that!"

The abrupt change in the conversation caught me off guard. The realization was that cataclysmic as she made the discovery — in her own mind — that it did apply to her little farm. Absolutely, and the sooner we begin thinking about how much of this whole approach applies to us instead of how much doesn't or can't or shouldn't, the sooner we'll make the kind of positive changes necessary to put us on the road to recovery: economically, emotionally, and environmentally.

The fourth advantage of relationship marketing is life-style. I know the world is full of high rollers. Some people seem to thrive on being big. Dad used to call it "being born with a big auger."

But for most of us, the price of size gets paid in ulcers, loss of our children to things we don't have time to teach them not to do, marriage breakups and a host of other life-style traumas. As the empire builds, so does the pressure. Financial pressure, the work load, dealing with employees — all of these take their toll on our life-styles.

We all know that money doesn't buy happiness. Some of the most unhappy people in the world are millionaires. Wealthy people often worry more about keeping their millions than the rest of us do about acquiring money.

I think one of the biggest differences between the pressures I encounter as a small potato and the pressures encountered by big potatoes is the amount of control we have over the situations that cause pressure. Big pressure for us is if a deer runs through the cross fence and the cows get into the next paddock. This is a far cry from labor relations trouble among the employees, or inability to finance the Friday payroll check, or a truckers' strike that keeps me from being able to get product to customers. These problems, because they result from interaction with people and other businesses, are much harder to control. They can't be "fixed" as easily as things under my direct supervision on our own farm.

This is the element that has led financially troubled farmers to suicide: when they realize that the financial picture is completely out of their control. As long as we are in the driver's seat, we can take all sorts of rough driving. But when we have no control over the vehicle and some yo-yo is behind the wheel and there's no way to escape — that's when life-style goes to the dogs.

No one can escape from the pressures of life, whether they be financial, emotional, physical or spiritual. But the chances of affecting those pressures, of dealing with them, of solving those problems, makes the difference between an enjoyable life-style and a terrible life-style.

We all have 24 hours in a day. Within the parameters of our gifts and abilities, we can choose how we will spend those hours. I have no desire to complicate my life with an empire. I'd love to see thousands of independent farms serving their own local clientele. That spreads more happiness than a huge company with a lot of employees.

The fifth advantage of relationship marketing is a term I'll call balance. It helps to equalize the relationship between producer and consumer. Certainly the producer must cater to the consumer; after all, the first rule of business is that the customer is always right. But in today's climate of liability, suspicion and selfishness, it is nice to be able to close the door on someone's face when that person is not a good patron.

Every public establishment has that list of people they call "troublemakers." These are the customers who cause store personnel to roll their eyes at each other, silently communicating the message, "Oh, boy, not ***him*** again."

Some sales can actually cost the establishment money, for a negative gross margin. Every sale is not a profitable one.

One of the best experiences I've had to illustrate this point was a woman who bought some beef from us and complained that it was so tough she finally had to put it in the blender, but it was still too tough to eat. Other customers who had gotten other quarters from the same animal said it was wonderful. In fact, one customer was in the process of enjoying her grilled steaks even as we talked on the telephone. This complaining customer loved our chickens, though, and when she came to pick them up I carried the cooler out to her car.

When she opened the trunk, it was half full of cases of Dr. Pepper. Being biological farmers, of course, we generally don't drink soda, and she knew this. She was embarrassed at what I saw, and immediately proceeded to offer the following explanation in her whiny voice: "I got some Dr. Pepper from the store last week and it was a little flat. I took it back and complained and they gave me all this extra to help compensate." She is the same person who looks at sweet corn and squeezes half the kernels before selecting the ears ... of corn she didn't squeeze. She also complained about the chickens; some minor detail.

The point is that she wasn't good for business. She was a liability customer. We knew our beef was good, our chicken was good, and our vegetables were good. I'll bet the Dr. Pepper was good too. We stood our ground and just got rid of her. She does not get an order blank any more. This helps to balance the producer-consumer relationship, so that we concentrate our efforts on profitable sales, appreciative customers, people who "get with the program."

Dealing only with people who share mutual appreciation is far more enjoyable than wondering what egghead is going to come through the door next.

This is not to say that we never make mistakes. We certainly have, and do. We've refunded money, given free merchandise, and all the conventional things that businesses do to keep customers. But we only do that when we are in error, not when there is some unfounded allegation. And we are free to concentrate our efforts on money-making folks. The others get deleted from our customer file.

All in all, the advantages of consumer education and loyalty, product quality, farmer life-style and sales balances make relationship marketing superior, in my view at least, to empire building. Life is too short, the family too precious, the farm too enjoyable, to sacrifice those things I hold dear for the mirage of empire grandeur.

These folks are our cheerleaders. Everybody ought to have cheerleaders — they are the essence of enthusiasm, good cheer and optimism. They stick right with you through thick and thin. Yes, I'll take our cheerleaders over the sale barn crowd any time of day.

It takes some time to build relationships. We've had many customers out for dinner. It makes you vulnerable to folks. But the rewards for ministering truth, information, mutual respect and appreciation — for building bridges — are worth every bit of the effort. The bridge of course is one that goes both ways, and when they need us as much as we need them the joy of farming could not be sweeter.

Chapter 39

Sales Versus Profits

Sales do not guarantee success. Marketing entails far more than transferring a commodity from your hand to a buyer's hand. Let's articulate the easiest, but costliest, sales for your business.

First is underpricing. Many beginning businesses fail because they want sales so badly that they underprice their product. In alternative agriculture, we generally offer a superior product. Our sales pitch might be that we have a more nutritious meat or milk product. Perhaps we'll say it's more environmentally friendly, or that by patronizing us, we stimulate our local economy rather than some distant region.

I know of hardly any businesses that begin because they offer a cheaper product. Certainly price is important, but that is not where the beginning enterprise can compete. Price wars need to be left to established businesses. The reason is that volume does generally reduce the price, and if we are trying to break into a market, our focus must initially be on supplying a superior product, not a cheaper one.

Dad used to admonish: "Remember, a business has to make a profit." Another of his favorite phrases was : "You may as well do nothing for nothing as something for nothing." The point was that when we are dying to sell, the temptation is to cut the price, even to the point of risking a profit margin. But especially in alternative food production, this decision brings us the wrong customers (folks looking for a "deal" instead of better quality) and sets us up for financial disaster when our profit margin per unit is not high enough. We want to offer a superior product at a competitive price.

Historically, farmers have notoriously underpriced their wares. In fact, consumers over the age of 50 typically think that if they purchase directly from the farmer, they should be able to get food at a 10-20 percent discount from retail prices. They remember the depression and post-Depression era when cash-starved farmers would greatly undersell retail outlets in order to get cash in their pockets. Of course, those farmers did not have the cash requirements of today's farmers, with personal property taxes, real estate taxes, car insurance, life insurance, medical insurance, fuel costs and on and on. They had some of these, but the ones they had were not as high as they are today. Taxes and cash requirements have gone up. Fortunately a new generation of American consumers is arriving on the scene that is more conscious of quality, ecology and regional sustainability (community).

If you begin marketing direct, don't be badgered into too-low prices by retirees, but stick by your guns. You know how much it costs to produce.

It's much better to sell 50 steers at a gross margin of $600 than 100 steers at a gross margin of $100. A well-managed small business can always grow into a profitable large business, but minimally profitable small businesses seldom grow into successful big businesses.

Because there is such a huge difference between wholesale and retail prices, you can usually have the best of both worlds: undersell the retail price and still get way more than you would wholesale at the sale barn. If we can show folks that they will actually save some money by buying in bulk, it makes those large quantities like quarters and halves easier to sell. The goal is not to just sell for a nickel or two over wholesale, but let the retail price be the standard and sell at or slightly below that level. And that is the retail level of equivalent cuts, not the retail level of sides and quarters.

It's important to know what is an equivalent price between a by-the-side sale or by-the-cut sale. The following represents a standard cutting procedure:

Retail Cuts	Percent of Carcass
Round steak	7.8
Rolled rump	2.8
Heel of round	1.6
Sirloin tip roast	2.9
Sirloin	5.3
T-bone and Porterhouse	5.4
Club steak	1.2
Standing rib roast	2.4
Rib steaks	4.3
Arm roast	3.6
Chuck roast	10.3
Boneless stew	2.4
Ground beef	21.6
Soup bones and misc.	7.1
Fat, trim, waste	21.3

This standard is quite close to our own experience. If you take these percentages and apply them to a half that dresses out, say, 250 pounds, you will find that you can charge between $2.10 and $2.20 per pound carcass weight before the price is equivalent to the by-the-cut price. This amount varies in different parts of the country. By pricing just a tad under the by-the-cut equivalent price, you can save folks almost $50 per half and this can become an excellent marketing point. The volume purchasing saves them in the long run.

Early on, establish a rewarding and satisfying gross margin and stick to it, building a customer base that appreciates your product for what it is, not just for what it costs.

"You get what you pay for." That is a phrase we use as a truism in our culture. We apply it to cars, clothing, stereos, appliances and vacation packages. There is only one thing exempted from it in our culture — food. Somehow we are supposed to be able to buy mediocre food, cheap food, and yet be healthy and vibrant.

The food industry should not reward mediocrity any more than the dry goods marketplace. It is time for non-food producers to be hit between the eyes with the notion that the same philosophy that produces shirts that fall apart, cars that stay in the shop, and cassette players that eat tapes also produces food that encourages obesity, fatigue, hospital stays and even death. The notion that some Twinkies and watered-down milk can provide the fuel to power a kindergartner through block-building, jump-rope and flash cards is ludicrous.

Farmers can no more cut corners and produce good food than consumers can eat pseudo-food and stay healthy. A lion's share of supermarket fare comes from the laboratory, not from the soil. It is concocted by an industrial complex that thrives on keeping people addicted to pseudo-food and pharmaceuticals. Perhaps no recent prophet saw this connection as early, or expressed it as eloquently, as Sir Albert Howard, writing in his classic, *An Agricultural Testament*, in 1943:

> "Artificial manures lead inevitably to artificial food, artificial animals, and finally to artificial men and women.
>
> "The ease with which crops can be grown with chemicals has made the correct utilization of wastes much more difficult. If a cheap substitute for humus exists why not use it? The answer is twofold. In the first place, chemicals can never be a substitute for humus because Nature has ordained that the soil must live and the mycorrhizal association must be an essential link in plant nutrition. In the second place, the use of such a substitute cannot be cheap because soil fertility — one of the most important assets of any country — is lost; because artificial plants, artificial animals and artificial men are unhealthy and can only be protected from the parasites, whose duty it is to remove them, by means of poison sprays, vaccines and serums and an expensive system of patent medicines, panel doctors, hospitals and so forth. When the fi-

> nance of crop production is considered together with that of the various social services which are needed to repair the consequences of an unsound agriculture, and when it is borne in mind that our greatest possession is a healthy, virile population, the cheapness of artificial manures disappears altogether. In the years to come chemical manures will be considered as one of the greatest follies of the industrial epoch. The teachings of the agricultural economists of this period will be dismissed as superficial."

How tragic that here we are, more than a half century later, contriving artificials more creatively than Howard ever could have imagined. And we have an infrastructure predicated on maintaining superficiality and what Charles Walters, Jr. calls "Neanderthal" thinking. The need of the hour is for you and me to individually touch folks in our neighborhoods, in our circles of influence, with the food-health link, with alternative thinking, with new paradigms that shatter decades of wrong thinking.

Folks will generally rise to our standards if we let them know what our standards are. The food industry is so busy trying to disparage the notion that beef can be good for you or bad for you, that eggs can reduce cholesterol or increase cholesterol, that the average person thinks beef is beef is beef. But when it takes twice as much ground meat than necessary to feed the family because half of it cooked off in grease, it is apparent that all beef is not the same.

Part of our marketing is to offer something so wonderfully different that it stimulates people to ask if there is that much difference in other things. Absolutely. Then we can steer them to the organic raspberry grower in the community, the organic apple orchard up the road, and so forth. But if all of us are busy trying to produce the same thing as the neighbor, where is the education, where is the difference?

If we provide folks with better food, then they need to understand that their part is to pay a premium. After all, if the ground beef goes twice as far as 89-cent material, are we not justified in getting twice the price? Actually, it should be more than that because of cooking efficiency, materials handling and packaging efficiency. The industry is running head-over-heels seeing how many corners they can cut, trying to fool folks into thinking chemicals won't end up in their groundwater, that reducing topsoil health is not dangerous. It is time to wake people up, to apply "you get what you pay for" to food.

The second pitfall of beginning businesses is what I call ancillary consumer services. Remember that the food giants do some things with incredible efficiency: transporting, packaging, and materials handling. Refrigeration comes cheaper by the square foot. Boxing and labeling is cheaper by the thousand. Transporting is cheaper by the ton. You get the drift.

We get letters all the time from people clear across the country wanting us to ship them chicken or beef. The nearest dry ice facility is 40 miles away. To ship 10 pounds of meat would cost us half a day's labor plus mileage. The cost would be exorbitant. No way.

We must appreciate that we cannot compete with the big operators at every level, and learn to stop our production or processing at the point where our quality/price enhancement can't compete with the conventional alternative. If a customer refuses to patronize your product unless he can get it "in the store" or on his front porch, he's not a customer who appreciates you anyway. Let him eat the conventional junk. Some folks wouldn't be satisfied unless you cooked it and spooned it into their mouths. Our desire to please must be held in check, because these ancillary services can bog us down in a quagmire of the most costly, least financially or emotionally rewarding aspects of the business.

Finally, keep accounts receivable low. Again, the temptation for a fledgling business is to extend credit because we are dying for a sale. But that sale doesn't do us any good until the money is in our pocket.

Dad's accounting clients were primarily agricultural and small businesses, and he always said that the single biggest factor in small business failure was too much money in accounts receivable.

We have always operated on a cash and carry basis. Beef customers pay cash before they can have their beef. We got taken one time on a quarter, and that was enough. We decided that if it couldn't be cash and carry, we'd shut down. Again, the hassle is as detrimental to your emotions as the financial loss is to your pocketbook.

Anyone not willing to patronize your product for this reason isn't worth having. Certainly there are extraordinary circumstances where a person forgot a checkbook or where a faithful old customer has a major car breakdown and is short of cash this month. You can extend some mercy in such situations.

But your policy should be to get cash on the barrel head. We used to require a deposit on beef orders, to insure that the customer would pick it up. Anyone not willing to participate can't be trusted; be glad you lost them before they took you for a ride.

You have invested in the product and brought it to a sale; it is not too much to require cash remuneration for your investment. I cannot emphasize enough the necessity of sticking to a no-credit policy from your customers.

There you have it. Set your prices so that no matter what your volume, your return is both emotionally and financially rewarding; steer clear of the temptation to do everything the customer wants; and let cash be your business byword. These rules will insure that your sales are beneficial sales, not negative sales.

Chapter 40

Developing a Clientele

We took a three-pronged approach to getting a patron base. The only problem with sharing this is that it may confine your thinking. Remember, this is our story. Yours may be completely different. But at least this may stimulate some creative thinking and you can make your own adaptations.

The first prong was samplers. Proctor and Gamble doesn't put shampoo in your mailbox for nothing. They know that the key to introducing new products is acquaintance. They also know that people tend to stay in their routine. How often have you gone down the cosmetics aisle and decided to try something new? It seldom happens.

We knew that the only way to get people to buy salad bar beef was to get it into their mouths. We gave samplers to anyone we thought might be interested. Over the years, we've never given anything away that didn't come back fourfold.

We've given meat, eggs, chicken, to salesmen, mechanics, roofers, product dealers, small businesses we patronize. People respond positively to a gift. A gift is always more appreciated than something you buy because it involves that emotional element of giving. The point is to get the product into as many households, onto as many tables, as possible. Samplers certainly work.

Although we've never done it, a spin-off of this would be to have a booth at a food fair or exposition, offering free pieces to folks who come by. Free samples are one of the underpinnings of successful marketing. We found a tremendous prejudice to non-grain beef. People by and large just knew it would be tough, stringy and gamey. To overcome that, we had to introduce them to it without any risk. The response has always been tremendous to this technique.

The second prong was education. We put together a slide program about our farm, titled it "Environmentally Enhancing Agriculture" or whatever the group wanted to call it, and began making presentations for local organizations. Every community has its Rotary Club, Ruritan Clubs, Kiwanis, Exchange Club, Key Club, Women's Club, Junior Women's Club, Jaycees, Garden Clubs, Retired Teachers Association, and American Association of Retired Persons (AARP).

These groups meet at least monthly and have program committees. By the time these groups have heard the United Way chairman present the fund-drive update, the volunteer fire department's report and the local school superintendent pleading for more money (that's a pet peeve of mine since we homeschool — bricks and mortar do not equal education) and the hospital administrator has explained the new high-tech tool they bought, these folks are ready for a different, interesting program.

Not only that, but these folks are bombarded every evening on TV by news reports explaining how environmentally-damaging farmers are, and how beef isn't fit to eat, and how animals are treated inhumanely. They really do not want that great guilt complex to sweep over them every time they eat a hamburger, but they just can't help themselves. They've been told it's all bad, bad, bad.

People hate to believe, deep down in their souls, that farmers are their enemies, but they have no choice. Unfortunately, many farmers are their enemies, but that only makes the situation worse. My slide program offered an alternative. I showed green hillsides, contented cows, shed-protected manure in the winter time, large-scale composting, clean water, healthy food — these folks go bananas over this.

It is what they've been yearning to hear in their heart of hearts for lo these many years. Finally they see that cows are not inherently evil, but it's how they're managed that makes the difference. The response to this program has been overwhelming. In fact, many of these groups have had me back a second time. It is like a breath of fresh air and it's interesting; it's different; it's dynamic.

AARP groups always offer door prizes to get their folks to stay after the meal instead of going home and taking a nap. I take some eggs to add into the door prize box and they eat it up. It shows I'm in the spirit of the party and they warm to the event. If it's a women's group, I usually take a steak in a slow cooker. After the presentation I'll take it out, put it on a platter, cut it up and let everyone have some bites. That way they can see how little fat is in it, taste it, handle it. Women's groups are always far more receptive to my program than are men. I think it's because they are usually the ones who take the children to the doctor, who stay up with them when they have a cough, and who have a more tender heart toward creation.

Men tend to sit cross-armed and demand double-blind empirical data before they will believe your point. Women have that intuitive sense that "this is right." Men say: "Prove it." I'll speak to a women's group ten to one over a men's group.

The program is educational, not a sales pitch. But at the end, quite innocently, I'll say: "Now if any of you would like to participate in this type of agriculture, I happen to have some order blanks with me and you're welcome to sign up."

Another advantage to this approach was the chance to defend it, to promote it, personally. We found early on that as soon as someone heard "organic farmer," they pictured a hippie wearing sandals, sporting a ponytail and coming in with hayseeds in his hair. Although this may seem unfair to many, I have found this stereotype to be very strong. Remember, a new McDonald's is still going up every day. A lot of people are still smoking, eating Lucky Charms for breakfast, and patronizing snack machines throughout the day.

When doing these presentations, I make the strong pitch for

changing purchasing habits. "If you don't like the degradation tonight's news will blame on farmers, huge food processors and payola inspectors, what are you going to do about it? If you patronize a food system that aids and abets these abuses, you are no better than the perpetrators. You cannot be an island and claim no responsibility for helping or hindering your own as well as the nation's collective health — environmentally, economically, and socially. What will you do?"

It is a strong message, and often leaves folks "stirred." The bottom line is that this educational, personally-representative approach is always good for a couple of new patrons. The whole idea is to take a proactive message right into the city and demand as much responsibility from them as I put on myself. What folks need is information, and education meets that need.

The third prong was to turn our patrons into evangelists. Our culture is starved for appreciation, and we let our folks know that we appreciated them spreading the word about us.

When Jane Doe would call and ask to be put on our customer list, we would ask her where she heard about us. "Oh, Mary Smith had us over to dinner the other night and served us some of your beef. It was so delicious I asked her where she got it and she told me about you."

We make a note of it and the next time Mary Smith picks up something, we tell her how much we appreciate her spreading the word about us and give her a chicken if she's a beef customer and a steak or roast if she's a chicken customer. Of course, if she's both then we give her some pork or eggs — you want to take advantage of these opportunities to "spread the acquaintance."

Just imagine if the next time you want into the hardware store where you do business the proprietor walked over to you and said: "Jill, you're such a good customer, how about just walking over to that tool board and pick anything you want. Take it home with you, our compliments." After you picked your lower jaw up off the floor, you'd pick up your tool and exit in a complete charismatic ecstasy.

"Guess what! You won't believe what happened to me down

at the corner hardware store! You ought to go down there and see this place!" You'd shout to everyone who came along. Yes, our world is starved for some appreciation.

Recently a fellow came by to discuss some business items and he told me about something he witnessed here last year that left a big impression on him. "I drove up and saw a lady park her car, get out and come up to you and give you a big hug. You don't get that at the supermarket." He is right, of course.

Listen, folks want appreciation. They want to know they can make a difference. They want to be a part, not just customer number 1,223,443,761. The two things supermarkets cannot do is provide high quality food and offer a relationship. They can transport and handle material, for sure, but they can't do what we can do with a closeknit group of like-minded folks.

The beauty of this approach is that it ensures high quality patrons. Existing patrons screen potential ones, and because many of these introductions are in social situations it insures that similar type people will be the new patrons. This is a wonderful screening procedure, and is better than any insurance policy money can buy. Our folks introduce our products only to folks they think will be good for us to have. It's a mutually-beneficial approach. Routinely our patrons ask for additional order blanks to give to friends, business cards to pass out, or for extra product to share with others.

Because salad bar beef is such a specialty, it is generally served in special situations — when "company" comes. This means it is normal for others to be introduced to our product.

The threefold approach we used to develop our patron base, then, hinged on samplers, education and evangelism. While this is certainly an unconventional advertising program, it is important to stay consistent with an unconventional product. Our experience, as well as that of others, shows that advertising an unconventional product conventionally never pays off.

That route brings people who are not aware of the program, who come for a number of wrong reasons. You can do a lot better with 100 patrons who are "with it" than 500 who are on today and off tomorrow. A loyal, supportive corc of cheerleader patrons makes farming enter a dimension that is almost inexpressible. When I talk about this aspect, it's easy to get choked up. I can't imagine trying to farm without the encouragement, counsel and strength of our folks. We love them.

A friend in Indiana began raising pastured poultry a couple years ago in addition to their stocker calves, which they'd raised for some 30 years. At the last pickup day, a customer asked him if he'd made a good profit.

Somewhat embarrassed (who ever cares if the farmer makes a profit?) he said he had. She immediately responded that she needed him to stay in business and he needed to make enough to enjoy it. "So you just raise your price if you need to because I need you," she finished.

He couldn't believe his ears. "I've been selling cattle for 30 years and nobody ever asked me if I made a profit, let alone tell me they'd be happy to pay more if I needed it," he said. The whole idea of relationship marketing, a loyal clientele, what I call "cheerleader" patrons, adds a dimension to farming that revolutionizes it financially, emotionally and spiritually. I highly recommend it.

Chapter 41

Communicating With Customers

Farmers routinely ask me how we keep 400 customers together and how we pre-sell everything. It seems inconceivable to many that our folks order chicken, beef, rabbit and pork once a year. Obviously that is quite different than going down to the local supermarket and picking up something for tonight's dinner.

We communicate via an annual newsletter and order blank, with enclosed self-addressed envelope. Initially, we used a newsletter and asked people to call. That did ***not*** work. You know how it is when you ask someone to call.

Then we went to the order blank format, but that didn't work either. Finding, addressing and mailing an envelope was too much work for many people. Obviously, many folks would do it, but too many did not.

Finally, we enclosed a self-addressed envelope and that cured everything. Now all the customer has to do is fill out the order blank, stick it in the envelope and put on a stamp, and drop it in the mail. To encourage quick responses, we operate strictly on a first come-first served basis. In fact, some customers now call us as soon as they get the newsletter, telling us what they want and following it up with their order blank. Some folks have gotten downright paranoid about missing out on something. I suppose in a couple of years we'll go to e-mail with our order blank.

Since our supplies and capacity are limited, we take the orders as they come in, dating each one so that we have a record in case someone can't believe they didn't send their order in soon enough to get on the first batch of chickens, for example. Imagine explaining to someone that even though we got their beef order before the end of March, all the split quarters are already sold (for October).

When everything is ordered, we make what we call our "safety valve" list. We go to that if we have a no-show order, or if we have extras. The overflow folks are wonderful to take the pressure off when you have extras to sell.

Our newsletter is the anchor of our customer base. We generally send it out in late February. It is primarily educational, since that is the ingredient in agriculture most consumers are starved for. We level with customers about our victories and our failures, our needs and our dreams. It is a wonderful vehicle to mass-communicate the who, what, when, where, why and how of our farm.

The key to a successful newsletter, I think, is to not put on airs. If you shoot straight, folks will respond and appreciate your candor. If you had some tough meat, explain why and what you are doing to fix it. If you are changing slaughterhouses, explain why. If you are increasing prices, explain why. Never increase prices without an explanation. Most people have no idea what it costs to run a farm, or how their lobbying for more schools increased your property taxes $500. This is all part of the educational dialogue that will underpin any substantial progressive steps in the rural/urban conflict.

Try to eliminate surprises and bring the folks into your living room, letting them see into your farm. They will love you for it. For many, you are the only farmer they know personally. That is a tragedy, but also a wonderful opportunity to think that you could actually mold and shape the view of farming, of land stewardship, of food production, held by a couple of hundred people. What a fantastic ministry you have, and it needs to be utilized to the fullest potential.

If you need cash, explain why and ask for it. If you have an idea, share it and ask for a response. Follow up next year, explaining the results of the question and what you are doing as a result. For example, a couple of folks suggested several years ago that they wanted to know other customers who lived in their area so they could carpool. We added a question in the order blank the following spring, explaining that anyone who checked the box would be put on a master carpool list which we would then mail to everyone on that list. This protected the privacy of folks who were not interested.

We did not want to violate confidence by publicizing our customer list. As it turned out, only about ten folks were interested. Most people enjoyed coming out, and didn't want to lose that opportunity. We mailed the ten names like we said we would, and addressed the issue the following year so that everyone would know we did heed the suggestion and followed through.

Specifically explain why your product is superior to what is available conventionally. We've printed lists comparing our products with supermarket fare, and that has always elicited an excellent response. When you get the differences articulated, side-by-side, it does make a graphic and impressive comparison. It's hard to believe how creative we've been at developing improper models.

People are always impressed with openness and candor. People always warm up to someone who risks vulnerability to bring us into their world, their plans, and their needs. Our folks have come to anticipate the newsletter like the harbinger of spring and summer, and even begin calling when they think it ought to arrive.

The order blank articulates the entire season's offerings of chicken, beef, rabbit, pork and whatever else we have. Beef, available in October, is divied up into split halves, halves, wholes and ground quarters (cull cows) with their prices and a weight range. Customers can check "Do you want liver?" so that no one is forced to have it. Because the money for beef can involve hundreds of dollars per order, it is imperative that people know a price range for what they are ordering. They can mark whether they want a larger animal or smaller one.

For most folks, this is the first time they have ever spent this much money on any food item at one time, except maybe for a wedding reception. You must do everything possible to make sure they understand that when they order a quarter of beef, they will be spending as much as $300, or a half, $600. It is better to talk them through it, or even have them back out, at ordering time than to have them faint in your arms at pickup time.

We also ask for comments and suggestion on the order blank so that folks can articulate their ideas about how we could improve something, or their praises and testimonials for what a wonderful job we are doing. Believe me, those praises are invaluable on a cold early March day when last year's money is almost gone. That is one of the unquantifiable emotional rewards of direct marketing.

Each customer's address and phone number are on a 4x6 card in our alphabetized customer file. We put the season and year on the card when we put it in the file so we know how long folks have been with us. Anyone who does not order in one year is purged from the file the following year.

Mailing is too expensive to carry non-buyers. It's funny when people call begging: "Please put me back in your box." A small direct marketing business needs to be lean; you can't afford to deal with customers who dillydally around. That is a tough, tough purge policy, but it means we have an active customer file.

This marketing technique gives us the luxury of growing for the market. Fall beef is ordered in the spring, giving us the entire summer to drum up more customers if necessary. Normally we are short, but we would rather be short on animals than short on customers.

Sometimes it takes a couple of years to whip a new customer into shape, but once he's missed out on something, he usually "gets with the program" the next year. We try to level with our folks, to provide them with a consistent high quality animal product, and require equal responsibility from them to plan their year's needs and order it.

If a split half is too much for someone, we encourage them to find a neighbor or friend to go in with them. Don't let them just go away; turn their reluctance into a new customer. What we find is that when people purchase a volume at a time, they tend to eat far more than they would otherwise. When that beef is sitting in their freezer, bought and paid for, they soon forget the expense and go ahead with consumption.

If you can get someone to purchase their first split half, they will usually eat it in a year even though they assure you that "we just don't eat that much beef." They will when it's sitting in their freezer, and they will when they don't have guilt pangs when they eat it. You've got a lot going in your favor once it's in their freezer.

By constantly purging and adding, the quality of our customers increases annually with the quality and refinements we impose on the farm. It truly is a wonderful marriage of rural and urban, consumer and producer. Communication is the key.

Chapter 42

Processing and Regulations

Salad bar beef is by far the best when it is processed in the fall, preferably after the first frost and before grass quits growing for winter. In the natural cycle of things, the animals "tank up" right before winter. The frost kills flies and the cooler temperatures increase the beeves' comfort to their highest level in the whole year.

From a marketing standpoint, seasonal production is always a problem but that is something to which the customers must adjust. As soon as the production runs counter to the season, palatability declines and we begin circumventing the system with things that do not make economic or ecological sense. Product integrity requires no grain and processing at the season's comfort peak.

Perhaps in high elevation arid areas on irrigated pastures (we won't discuss the ecology of irrigation — it can be acceptable or not acceptable) this timetable could be fudged a little. In Argentina they fudge by grazing corn (maize) and other annuals. The same thing could probably be done here; and certainly this system is in its infancy but coming on fast. In keeping with my philosophy on perennial polyculture and no tillage for beef, however, I do not endorse such techniques. At the same time, I do not have strong enough notions about it to argue. As long as we understand the pitfalls of such techniques — tillage, risky annuals, monocultures — I'm certainly open to experimentation.

I am not qualified to discuss such programs and I will leave that to others. I do think that strip-planting and offering the cattle a cross-section of monocultures can alleviate some of the monoculture problems. Extremely long rotations between heavy feeder crops can also alleviate the soil-deteriorating properties generally inherent in things like corn. Using subterranean clover, building up organic matter through longer pasture sequences or planting into smother crop mulches can reduce risk and encourage water cycling in an otherwise anti-water row crop program.

Some practitioners show great success with such programs, and my hat is off to them. I suggest, however, that these are all refinements to be made ***after*** the effective management-intensive grazing, nutrient cycling and water systems are in place. Our love affair with new plant varieties and equipment usually causes us to get the proverbial cart before the horse. But to try some of these techniques to spread the window of palatability for salad bar beef is certainly commendable as long as the major components are fully refined and functioning.

Most importantly we want to harvest the animals when they are the most comfortable and on the highest quality forage. That will insure the most succulent beef.

Ideally the animals are actually slaughtered in familiar surroundings. This eliminates the stress of transport, unloading and strange surroundings. But unfortunately this is not acceptable except for personal consumption. Although there may be special ethnic exceptions, generally regulations prohibit selling beef slaughtered and processed on the farm without some inspection.

Before discussing inspection, let me make the point abundantly clear that inspection has nothing to do with food safety. That may sound like a shock, but it is true. Upton Sinclair's *Jungle* notwithstanding, inspection grew out of a demand by people too far removed from their food who wanted salvation by legislation. While I would never condone filthy slaughtering practices — along with poor employee treatment — I would not run to the government for answers.

Most of the things the government gets involved in, it either louses up or turns the whole issue into a political game in which the small players always lose. Parochial schools and home schools are constantly assailed by bureaucrats intent on maintaining the government agenda. Performance of alternatively-educated students has ***never*** been an issue; aptitude test performance, on average, is ***always*** well ahead of government-educated students. The issue is not quality, but access to alternative agendas.

Once the government regulated health care, homeopaths, which accounted for 50 percent of doctors operating in 1900, became villains and quacks because the techno-medical community that opted for cut-and-burn systems cranked out both doctors and regulators. Anyone who thinks the Food and Drug Administration and the big pharmaceutical companies are not in bed together has simply not looked at the facts. The same can be said of the USDA and the multinational giants like Archer Daniels Midland, Monsanto and Ciba-Geigy.

Labor laws are skewed toward big business. If a company operates a bus to pick up workers, that is a tax-deductible business expense. But if a worker drives all the way to work at a small company, none of that commute is deductible as a business expense. The list could go on and on, but the fact is that regulators and big business are in cahoots and what started out as a sincere desire to "protect the public welfare" quickly turns into a political game that pits the "insiders" against the "outsiders" and soon transforms market access into a payola program.

I can butcher a beef in the backyard and give away the meat to a hundred people through a volunteer fireman's carnival fundraiser. But if one of those folks comes out and asks to ***buy*** a pound of hamburger, I dare not sell it uninspected or I'm a criminal. Clearly if public safety were the issue, it would be as wrong to ***give*** the meat away as it is to ***sell*** it. Plenty of people have tried alternative schemes to selling, but government attorneys, operating at taxpayers' expense and at the behest of the large moneyed interests, find themselves facing reprimands if they do not enforce the letter of the law, regardless of how unreasonable it may be.

It is amazing that people who complain about how the government handles things turn right around and demand that the government protect their food. This is schizophrenic reasoning.

Freedom and risk go hand in hand. When we eliminate risk, we also deny freedom. Freedom to choose cannot guarantee proper choices. Inherent within a freedom to make a choice is the latitude to make both correct and incorrect choices. Under the guise of "protecting" people from making improper choices, we have denied folks the ability to make wise, alternative choices.

If the only meat allowed on the market comes through $100,000 slaughter houses, then I have no choice to purchase meat from a neighbor who dresses a beef in the fall on his farm. This is the great evil of regulation: in denying the opportunity to make a bad choice ("get burned" so to speak) it inherently precludes the freedom to make a positive alternative choice. It cannot be both ways, and this is the great flaw in thinking we can stop abuse by regulating it. Freedom to make good, alternative choices ***requires*** that we be free to make improper choices, to "get burned" once in awhile. It cannot be both ways.

Nothing substitutes for personal knowledge, personal relationship, and personal responsibility. As long as people demand that someone else be responsible for their food safety, those same people give up their right to find better food. Their food will always be defined by the parameters of their protectors, and their protectors are wined and dined by those able to do so, by the elite who too often care not so much about health and quality, but about fattening dividend payments to stockholders.

I am not an anarchist by any means, but having been through the bureaucratic mill on occasion, I stand unequivocally on the premise that the best thing we could do for food safety, food quality, and food creativity is to reduce food safety laws and unleash the entrepreneurial spirit that is alive and well in America if only it were not shackled in dungeons of red tape. When millions of animals are slaughtered around the clock in a facility, filth is difficult to control.

But when it is done a few days a year in a backyard, the sun, breeze and break between processing provide better sanitation than all the toxic chemical bactericides in the world. Furthermore, small operators who cannot afford liability insurance, who do not have a bunch of Philadelphia lawyers on retainer, tend to be far more careful about cleanliness and quality than the big operators.

From a factual, philosophical and emotional standpoint, food safety laws stand in the way of progress. Exemptions should be granted for small producer-operators, to encourage rural enterprises and bioregional food sufficiency. These laws currently aid and abet centralized agriculture, distance between producer and consumer, high-calorie requirements to bring food to the table. It is time for an overhaul, a balance, a recognition that we have gone too far. We have passed what was necessary and reasonable.

That issue is interesting because it points up another basic flaw of regulations: the appetite is never satiated. Bureaucrats never seem to realize when they've gone far enough. They're always busy writing one more regulation. Can you imagine an agency telling all its workers one day: "Well, folks, we're pleased to announce that all our regulations are working and we don't need anymore. So effective Friday, everyone involved in writing new rules will be laid off and we'll just spend our time enforcing what we've got." Wouldn't that be something?

Here's a question worth pondering: "What if all the energy and time expended by consumer advocacy, environmental, and animal rights groups to get increased "protective" regulations, had been expended instead on channeling their constituents to patronize businesses offering environment-friendly, consumer-friendly, health-friendly products?" These groups are huge, and if they would unleash their constituencies on the marketplace instead of on Congress, I submit their agenda would be far more advanced than it is.

The problem is that such thinking requires adjusting life-style and spending habits. It requires fundamental changes in the way we live and act. Perhaps no single story better illustrates this than an experience I had in Washington D.C. a couple of years ago. I was asked to participate in a high-level discussion between a prominent sustainable agriculture think tank, the USDA and the Department of Health and Human Services. The day-long meeting got underway and after the initial niceties were exchanged, the floor was thrown open for input.

Being the only farmer in attendance (they supposedly wanted one there for balanced perspective) I put in my thoughts about reducing government regulation on small farmers who wanted to sell items to their neighbors, rather than starting another government program duly funded with tax monies. A lively discussion ensued and it was apparent that they expected their "farmer" representative to be seen and not heard. "We've never had a farmer say anything before," they said.

The upshot of the day's discussion came from the big sustainable agriculture organization's paid lobbyist, who stared at me incredulously from across the oval table and said: "Listen, I don't have time to get in touch with my food supply." That jolted me. Here I was in supposedly friendly territory, talking to someone partly paid by my dues and subscriptions to protect my interests, and she was that far away from thinking about community, about small business, about freedom of choice and neighborhood economies. As the Virginia Slims commercials say: "We've come a long way, baby."

It will be a long road home, but each of us, one person at a time, marketing to our neighbors instead of foreign countries, can begin building bridges instead of barriers. Our government puts an incredible amount of time and energy in exporting our farm products all over the world, but puts roadblock after roadblock in front of us if we try to sell a pound of hamburger to a neighbor, a pound of sausage to a relative, or a pound cake to a friend at church.

With that said, let's discuss what we ***can*** do. Most states have three levels of inspection: federal, state and custom.

Federal inspection allows the meat to be sold anywhere in the country, across state lines. An inspector is on the kill floor at all times.

State inspection allows meat to be sold anywhere in the state: it cannot cross state lines. An inspector is on the kill floor at all times.

Custom inspection does not allow meat to be sold after slaughter and does not have an inspector on the floor. This level arose primarily for farmers who wanted their own animals for personal consumption. Farmer Jones takes his steer in and brings it home in packages a couple of weeks later.

Each of these levels carries a different set of regulations. For example, under federal and state inspection the facility must provide a bathroom and office for the inspector, bathrooms for employees, etc. The kill floor ceiling must be a little higher than the one in a custom facility.

Of course, the slaughtering price escalates as we move from custom to federal. Generally federal costs about 10-25 cents more per pound of carcass weight than custom. This adds a substantial cost to processing. The difference is that custom-processed beef is marked ***NOT FOR SALE***, while the state and federal packages can be sold from your freezer. Furthermore, restaurants, because they are selling meat after slaughter, cannot use anything but federal or state-inspected material.

For decades, the custom facilities were used by farmers in the freezer beef trade as long as the beeves were contracted for prior to slaughter. In other words, ear tag number 32 is owned by Jane Doe (right side) and Jim Smith (left side). The price could be based on carcass weight and the prearrangement showed involvement between producer and buyer prior to slaughter. This requirement differentiated between a sale arranged while the animal was still alive and one transacted at a supermarket meat counter where the buyer had no original knowledge about the animal and it was an after-the-fact-of-slaughter deal.

With the increased number of farmers using custom facilities, however, large packers pushed through a new interpretation of the law so that now these animals must be sold by the head or by liveweight regardless of pre-slaughter arrangements. In other words, any price based on dead weight is deemed an after-slaughter sale and therefore illegal in a ***NOT FOR SALE*** custom facility. Of course, this is ludicrous inasmuch as things are routinely sold based on post-dated conditions. Houses are sold before occupancy all the time. In fact, bureaucrats have assured me they have no idea if this new interpretation will hold up in court.

The compromise we've worked out, however, is to sell animals for $1 per head and then add on shipping and handling charges based on carcass weight. This is the way we make up the invoices and it is perfectly legal. There are plenty of ways to skin a cat, and it is time for us entrepreneurs to show the same savvy as our pioneer forebears who went bravely into uncharted territories. Why would anyone want to sell to their neighbor when they can just go down to the sale barn and forget the "people" hassle? Well, some of us are just crazy enough to want to sell to our neighbors.

Custom facilities are by far the smallest and cheapest. The single biggest problem in meat slaughtering is integrity. Beef is notoriously easy to steal because it is not counted. Who will miss a pound? There is no way to check it. I encourage folks to use several facilities and then settle for the one that does the most conscientious, dependable job. We've had some absolutely terrible experiences, so be forewarned that there are good facilities and unscrupulous ones.

One of the best ways to check is to take a couple of animals and give the same cutting instructions. Weigh the take-home packages and compare that to the hanging weight. It should be consistent from one processor to another. If it is different, use the processor that yields the greatest take-home percentage. He's least apt to be pilfering some T-bones on the side. The bigger the facility, the more fingers are in the pie and the more opportunity there is for loss.

To my knowledge, there is no alternative to state or federal inspection for restaurant sales. Also, if you want to sell the meat by the cut out of your freezer, it must be state or federal inspected.

We sell everything by split half, side or whole. By selling a split half we always come out even, rather than having many extra fronts relative to hinds. Our cull cows we sell as all ground, a quarter at a time.

The processor must allow the beef to hang for at least two weeks. This hanging period is critical for muscle tissue breakdown and tenderness of the meat. If he balks at this time period, go somewhere else or take the best you can get. You could soften him up with a couple of homemade pound cakes or cookies.

We do not ask the butcher to do anything special with our carcasses. He cuts them just like any grain-fattened animals. He keeps each person's portion separate on trays and shoves the buggies of trays in a walk-in freezer. Be sure your butcher uses an identification system that will stay on the tray. If you walk in the freezer and there are identification tags littering the floor, you do well to wonder if the beef you're getting back is really yours.

We have used both custom and federal facilities because we sell to both private folks and commercial establishments. We enjoy the custom because it is close, small, and impeccably honest. It is too bad that our county, which used to have a dozen, now only has three. Perhaps if more farmers would begin selling salad bar beef to their neighbors there would be a resurrection of these facilities and we could reduce the mileage under our steaks. Besides, these small plants do not need to mist carcasses down with chlorine, and that is a distinct advantage. Let's all do our part to bring our food production and processing system back home where it belongs.

Chapter 43

Pickup Day Logistics

Since the time span between ordering and when we actually take the beeves to the slaughterhouse is usually close to eight months, we send an instructional letter to each customer the day the beeves are arrive at the abattoir (slaughterhouse).

The instructional letter begins by reminding them of what they ordered and asks them to call immediately if that is in error. This is our verification of their order.

Then it tells them to call the abattoir and give cutting instructions. Split halves do not allow the same variety of options available when you offer front and hind quarters. Essentially the butcher cuts the side and makes two piles of every cut. There is scarcely a pound difference between the two stacks. The main variable is packaging size. The primary information the butcher needs to know is how many people are in the family. Then he packages accordingly.

Since moving to split halves, we have found no customer disappointment at losing some choices. Now everyone gets a piece of that hind, so everyone shares equally in the top cuts and bottom cuts. Selling fronts and hinds is a nightmare because no one wants the fronts, even when your price spread makes the hinds ridiculously expensive.

The other main item on the instruction sheet is the pickup date. We offer two mornings: a weekday and a Saturday, 9 a.m.-

noon. All customers must arrive at the abattoir that day to pick up their meat. We talk straight to our folks about the necessity of showing up during those time periods. If they miss it, we tack on a hefty surcharge, and if they complain about the extra charge, we cull them from the customer list. We take responsibility for producing the best, cleanest salad bar beef in the world, and we ask them to exercise equal responsibility in showing up to get it.

These are extremely busy mornings, but it is amazing how much beef you can move in a few hours. The folks show up and I make out invoices for them as they arrive. While they write a check, I go into the freezer and bring out their beef. We transfer it from the buggy tray into their cooler or box and away they go.

It's really quite simple, but it took us several years to uncomplicate it. Isn't that the way it always is? We start complex and gradually, if we're going the right direction, adopt a more elegant simplicity. The butcher does not want to handle our money or get in the middle. He doesn't even want to deal with our customers ("those city folks").

We pushed him hard in the beginning, forcing him out of his paradigm a little bit. Some chocolate chip cookies or pound cake can really help move folks. But now we are a big customer and he courts our business. He prepares an itemized statement of each person's beef and all the processing costs. We pay him one big check on the first pickup day so he gets his money right away. He does not have to handle a bunch of checks; he gets paid on time, and for him it is a nice volume in and out.

We do not need to answer technical butchering questions — he does. And he's good at it. He doesn't have to deal with our customers, which is what we do well. It's a perfect mating of our separate gifts and works well. Pickup days are hectic for a couple of hours, but that is our big payday. It is the last big interaction between us and our customers for the year.

We do see a few from time to time throughout the winter, but for the most part we will not see these folks until the first batch of chickens the following spring. We are all wishing each other a happy winter and Merry Christmas and expressing thanks for a bountiful season. It's a festive atmosphere, combining good food, good fellowship and good community. What a wonderful way to make a living.

Extending the Salad Bar

Chapter 44

Vision

What would the beef industry look like if salad bar beef became the new paradigm?

First of all, the power of the grain cartel would be all but eliminated. I've read article after article about the evils, the manipulative power, and the tyranny of these huge trading entities. I speak at farm conferences and watch overworked, underpaid farmers wring their hands and shake their fists at these companies.

I do not think they are necessarily out to destroy anyone. They are made up of people who put their britches on in the morning one leg at a time, whose children play in little league and who go to church and play golf just like anyone else. The problem is the power they wield and the seemingly helplessness farmers feel in the face of that power.

But if farmers would opt out, one at a time, and then in droves, by the thousands, and quit producing the corn that goes to the elevator to be bought back by farmers to produce beef, it would cut off the supply of goods to these conglomerates. Then where would their power be? Then the temptation would be for farmers to join bigger cattle buying entities. They would have to eschew that temptation as well.

But just imagine if 70 percent of all the grain grown were returned to perennial polyculture, managed intensively to raise salad bar beef. What a different landscape it would be. Perennial polycultures, or salad bars, do not require tractors, combines, silos, plows, discs and spray equipment. Large farms would be obsolete and as they downsized there would be unimaginable opportunity for displaced factory workers and grain elevator workers.

Erosion would be all but eliminated and the rivers would run clean again. The money would be so good from salad bar beef production that in order to get people to raise grain, the price would have to double or triple. Corn would go to $6 or $8 a bushel. That would force chicken prices up into the $1.50 a pound range and suddenly beef and chicken would be comparably priced.

The only reason chicken holds such an advantage over beef is because two pounds of corn will produce a pound of chicken but it takes seven pounds of corn to make a pound of beef. Salad bar beef and expensive grain would equalize the price and deny poultry its cheap grain advantage. This would, in turn, greatly increase beef consumption.

Vegetarians and environmentalists could not accuse cattle producers of evil practices and many of them would undoubtedly join the ranks of beef eaters. That would further increase demand and hold prices up as the amount of beef produced doubled and tripled in the nation, eliminating use of petroleum fertilizers because of all the manure produced.

Because beef would not need to be trucked to feedlots in the grain regions, states would keep their own beef dollars at home. Backgrounding states that currently lose millions of dollars a year by exporting $400 calves and importing boxed beef at $1,200 a head would keep all those dollars at home. New community processing facilities would spring up, revitalizing rural economies.

Cattle would be freed from the bondage of cross-country trips in potbelly trucks. They would be unstressed and allowed to enjoy fresh air and sunshine, green grass and soft soil. Baby calves would be born in late spring among dandelions and clover instead of in January on ice and snow and musty hay.

Through relationship marketing thousands and thousands of farmers would develop a loyal nucleus of customers who would provide financial assistance, part-time labor and emotional support. Beef consumers would learn about true land stewardship from farmers instead of through the jaundiced eye of office-building-sequestered eggheads. A new dynamic would emerge between rural and urban communities, a new understanding and appreciation.

Feedlots would cease to exist, as would their smells and dust. Diesel fuel and farm machinery could be converted to better uses and conservation/recycling efforts as cattle harvested grass directly. The quality of life on the farm would be almost heavenly as tillage, harvesting and chemical application became obsolete.

City folks could enjoy beef without guilty consciences. Because the beef would be lean and lower in cholesterol and saturated fat, they could spend less time at the doctor's office, less time worrying about diet, and more time reading to the children or attending family reunions.

Yes, it is easy to dream about such a world. I am not so naive as to think this will actually happen. But it can happen in the lives of individual farmers in their communities. Just because it seems laughable to imagine what a salad bar beef paradigm would do to our culture does not mean that individuals should not pursue such a goal. All of us, in whatever capacity we can, should do our utmost to facilitate such a noble vision. It is right for our families, for our communities, for the land, and for the animals. It is right and it is good. Let's take the reins and pursue the truth.

Chapter 45

Cooking Salad Bar Beef

No book about salad bar beef would be complete without addressing the cooking issue. Because there is such prejudice toward non-grained beef, this topic is cultural and emotional. After all, when a societal norm is established, it has the weight of law.

Is venison tough? Is elk tough? No, of course not. Certainly we may have eaten some tough venison or elk, but to say that they are inherently tough is ludicrous. Many folks are familiar with tender, succulent, melt-in-your-mouth venison and elk. Was it grain-fed? No.

Clearly grain feeding is not ***necessary*** to produce tender meat.

Do we cook these differently than USDA prime? Probably. And therein lies the rub. Different products receive different cooking treatments. We would no sooner cook salad bar beef like fat beef than we would cook venison like fish. They're two different things.

Salad bar beef should be cooked slow. For your health, meat should always be cooked at low temperatures. One of the foremost pioneers in the natural foods field was Adelle Davis, whose book *Let's Cook it Right*, explains why meat should be cooked slowly.

At low temperature, meats shrink less and are more attractive, juicier and tasty. It takes less energy to cook it that way, which saves money. At low temperature, the connective tissues soften to gelatin, and all of this happens without careful hovering-over.

At high temperature, the proteins toughen. In addition, the outside may be burned while the inside remains raw, not a tasty combination for any meat lover. The protein contracts, shrivels and becomes hard, dry and tough. She says: "The moist heat of water or steam penetrates meat twice as quickly as the dry heat of the oven or broiler at the same temperature." For your health and nutrition, slow cooking of all meat is the proper way. And that includes steaks.

The great non-grain beef gourmets of the world cook it slow. That does not mean the meal must be long in preparation. Today's crock pots and slow cookers are wonderful. We use them all the time: plop the meat into the cooker in the morning and come back at suppertime. Put it on the table and the meat is ready. That's faster than grilling, microwaving or parboiling. And it is so tender it just falls apart.

For fast cooking, we highly recommend marinades. It is truly unfortunate that fast cooking has become an American tradition. One thing is for sure, that procedure is uniquely American. It certainly was not native to North America, nor is it currently practiced anywhere else in the world. Is it possible the whole world is crazy and we are the only ones that are right on this issue? I doubt it.

Here is a recipe that is one of our favorites, and it illustrates something that can be done with round steak besides fast cooking on a grill.

BAKED ROUND STEAK

2 pounds round steak 1 inch thick
½ clove of garlic

3 Tablespoons vegetable oil
1 (6 ounces) can tomato paste
½ teaspoon salt
1 small bay leaf
¼ teaspoon thyme
¼ teaspoon sugar
1 green pepper, cut in rings
salt and pepper
¼ cup flour
1½ cup water
1/8 teaspoon pepper
1 large onion, sliced

Cut steak into serving pieces, trim all fat. Rub with garlic, sprinkle with salt and pepper. Pound flour into steak. Heat oil in large skillet, brown steak on both sides. Remove meat and place in casserole. Pour off fat from skillet. Mix tomato paste, water and seasonings; heat in same skillet. Arrange onion and green pepper over meat in casserole. Pour tomato mixture over all. Cover tightly with lid or aluminum foil. Bake in moderate oven (350 degrees F.) for 1½ to 2 hours, or until meat is fork-tender.

It is also very good when put in a crockpot and cooked all day — about 7-8 hours.

Enjoying salad bar beef is not arduous. It is fun. If a gourmet chef thinks it's the best, that's good enough for us. Long, slow cooking enhances the natural rich flavor of salad bar beef. It is not tough and it is not gamey. It is the best beef in the world. But it is not fat beef. And because of that, it needs to be handled with the respect due its reputation. Such honor will yield truly outstanding results.

Appendices

Appendix A

Newsletters

January 1984

Attention friends of Polyface, Inc.:

Greetings from the farm of many faces. We trust your Christmas and New Year's memories are rewarding.

This is our first clientele newsletter. We hope it will be a periodic production to encourage communication and mutual satisfaction.

VEGETABLES: Last summer for the first time we sold our organic garden produce. As a result of several sundry sales of surplus veggies, we learned that people need this high quality produce. To expand this enterprise in 1984, we decided to offer cabbages, peppers and squash in quantity. Only if we have a surplus of other produce will we offer more variety. These three items, however, usually do well enough for us to feel good about promising them. We will order fungicide-free seed at the end of January to grow sets in our greenhouse for later transplanting. We need to know how many of these vegetables you think you might want over the course of the growing season. Prices will be whatever supermarkets are charging at the time.

Cabbages will be ready sometime between May 1 and June 30. Peppers and squash should be available from July 1 until frost.

POULTRY: Because of the tremendous response to the 400 almost-organic fryers we raised in 1983 we are looking at this item closely for

1984. But frankly, we had problems. Not with the production, but with the processing. We sold them for 65 cents a pound. After monitoring grocery store prices for awhile, we discovered radical fluctuations between 59 cents and 85 cents per pound. Not including our 50 man-hours of processing time, we netted only 50 cents per bird. Honestly, that's not enough.

Unless we find a way to net more profit we can't justify the time. We are examining other processing options to reduce its time. We are also convinced that we would have to charge 85 cents per pound dressed to make the project - which we enjoyed - worthwhile. If you would be willing to pay that price, please call us.

Right now, we aren't sure whether or not we will raise them. But since we probably won't be mailing another newsletter prior to our decision, we would like to know if there is enough interest to pursue poultry production. Then if we do decide to raise them again, we would call you who are interested. Although we fear boring you with these details, we would rather level with you about our problems than to, without explanation, either raise the price or refuse to raise the birds.

BEEF: We have three animals ready to process the first of June. Since our sales went well last year, we are changing some procedures. We will offer front quarters, hind quarters and halves at three different prices to conform to the meat trade. Beef commodity analysts disagree on what will happen this year so we aren't quoting prices yet. All we promise is that our prices for better beef will be lower than grocery store prices for smaller quantities.

Several of you have said that your beef isn't lasting like you thought it would. That is understandable with tender, organic, lean beef. If you would like to get another quarter in June to tide you over until October, when we hope to offer beef again, please call us. We will be glad to answer any questions, too.

We regret that because of a problem we had last fall, we must request a $25 down-payment on all beef orders. It is illegal for us to sell it after processing and this will, we believe, help to insure our orders.

So far, we have heard nothing but praise about the beef. If you have not been satisfied, please call us. Sometimes a cooking technique can make a world of difference. You who are satisfied, we appreciate your encouragement. If you are happy enough with the product, please mention us to your friends and neighbors. We hope that we will not have to send any more of our special animals through conventional wholesale-packer-supermarket channels.

RECREATION: A questionnaire that several of you filled out when you purchased beef revealed significant interest in camping, hiking, picnicking and fishing. We appreciate your candor. We are researching these areas and hope to make some positive reports by year's end. Progress comes slowly. Maintenance is what takes our time, isn't it?

FIREWOOD: Lord willing, we will also have plenty of firewood available during the summer and fall. We are cutting now and stockpiling it near the house.

At the risk of sounding condescending, we know how easy it is to delay. Before now, we have marketed our items by telephone and thereby received an immediate response. But we have an expanding customer and product list which preclude using the telephone exclusively. Truthfully, we fear everyone setting this newsletter aside and forgetting about it. Please don't.

As we close this letter, a milestone for us, we again extend heartfelt appreciation for your interest in our pursuits and biological farming methods. If you have never been here to see firsthand what we are doing with our little niche of God's creation, please arrange for a tour. We deeply appreciate God's blessings, viewing our endeavors as a ministry in every sense of the word - to consumers, agriculturalists and our society.

If you have any questions about any of these items, or if you have suggestions, and certainly (ha!) if you have orders, we are as close as your telephone or mailbox.

Expectantly yours,

Joel F. Salatin, Pres.

March 20, 1985

Dear supporters:

Happy Spring! We trust this newsletter finds you anticipating nature's awakening and marveling at the Creator's handiwork.

BEEF: We guarantee that ours hasn't eaten broiler litter, the new rage recipe. Neither is it eating antibiotics, which have been blamed recently for a death traced to a North Dakota feedlot.

The next processing date is early June. The price is the same as last year: front quarters, $1.28 per pound, hind quarters, $1.55 and halves, $1.35. The price includes cutting, wrapping and freezing. Hearts, livers, tongue, kidneys and bones are also available. Please place your order by April 15 since supplies are quite limited.

POULTRY: Because of the tremendous demand for our broilers, we invested in some snazzy processing equipment to accommodate all orders. Depending on demand, we plan three rotations to be ready about June 4, July 30 and Sept. 24.

These birds will weigh 3½-4½ pounds and the price is $1.00 per pound. We will dress and cool them to well water temperature.

FIREWOOD: Again, a tremendous demand has kept us cleaned out nearly all winter. Effective immediately, the price is $25 per pickup load. We cut it to length and pile it here at the house. You haul and split. Call before you come to make sure we have some down. It is sold on a first come, first served basis.

VEGETABLES: No demand for two years. If you want something, let us know, but we are not planning to plant for market this year.

RECREATION: We now offer on-farm vacations. Your accommodations are in the two-bedroom fully-equipped mobile home adjacent to our yard. Two meals per day are available. We would welcome you or your acquaintance to hike, camp, explore, learn about organic food production, talk with the animals and "get away from it all." Rates flex with your meal desires and length of stay.

SLIDE PROGRAM: We offer this 20-30 minute, fast-moving program to promote conservation, eco-farming, nutrition and healthy living as practiced by Polyface, Inc.

Please accept our deep appreciation for your encouragement and support. Call us today with your orders, questions, criticisms and to schedule an on-farm tour.

Expectantly yours,

Joel F. Salatin, pres.

Polyface Inc.

Spring 1986

Dear supporters,

Happy Spring! Please allow me to crank your thoughts ahead for a moment as we share together.

For your convenience we are enclosing a simple order form and self-addressed envelope with this newsletter. The order blank has plenty of room for you to write additional desires or clarification. After receiving your order, we always get back in touch with you, so don't worry about misleading us. We just want a rough idea of what you want. We hope this tool encourages responses, enabling us to plan schedules better.

Many of you have already placed orders -- if that is the case, your order blank is marked and all we need is a breakdown on how many when. If you're flexible, indicate that and it will be a big help to us.

Just a word about the ground beef. This is from animals we don't believe measure up to Polyface cuts quality (slower-growing animals or cows being culled from the herd for some reason). Depending on size, quarters may yield anywhere from 30-60 pounds. Those of you who have purchased these animals from us in the past know that the ground beef quality is absolutely superior. And, as with the rest of the beef, it's biologically grown. All prices are based on hanging weight and include cutting, wrapping and freezing.

Because our homegrown broilers are in such demand, we plan to add two batches to the rotation. Last year orders approached 1,000 and we hope to top that figure this year. Remember, you must order your birds before slaughter. The price is the same as last year, $1.00 per pound dressed.

Two other quick items. We hate junk mail, so if you want us to stop sending you this newsletter, tell us and we will gladly oblige. Secondly, feel free to introduce us to neighbors and friends. We give discounts to customers who introduce us to others who become customers.

By all means, communicate with us if you have questions, advice or encouragement. Someone is always here at one of our two phone numbers. Thank you for your patronage. Come and see us.

Joyfully at your service,

Joel F. Salatin, president

Polyface, Inc.

Fall 1986

Dear supporters,

If you've had as good a year as we have, you're on cloud 9. The cows and chickens have never looked better or been more contented. Truly we have much to be thankful for.

Fall beef will be available in early November. Some orders have already come in, so please respond soon while supplies last. One quick pitch. Forage-fed beef has less fat, and the fat that is there is UNSATURATED. Grain-fed beef and especially steroid-fed beef (nearly all commercially-produced beef) contains SATURATED fat due to altered metabolism. Seems like the more we learn, the more confidence we have in our special products.

The demand for our homegrown broilers is continuing to exceed our wildest expectations. We produced about 1200 this summer and have had to turn away many late-comers who asked for some after the last batch had been ordered. We only order chicks to fill the orders we've received, so there's a fairly healthy lag time (12 weeks at least) between your order and the finished product. We apologize to those of you who got a "no" late in the season, but trust you'll understand the situation and place your order early for next year.

By all means, communicate with us if you have any questions, advice or encouragement. Someone is usually here at one of our two phone numbers. Thank you for your patronage. Come and see us.

Joyfully at your service,

Joel F Salatin, president

Polyface, Inc.

Spring 1987

Dear friends,

What an exciting time to live. Never have the challenges been more challenging or the opportunities more stimulating.

We greatly appreciate the many of you who have already ordered broilers for May and June from our fall newsletter. The response was wonderful. If we already have an order from you, we've noted it on the enclosed order blank.

We are now taking orders for broilers that will be ready about July 18, Aug. 11, Sept. 5, Oct. 3 and Nov. 2. We've added the November batch to reduce the time until the May broilers. Remember, we operate on a 12-week lag, so we need orders now to know how many chicks to purchase, especially for July. You still have some time for the later batches.

It's time to order beef for June. Mark the order blank as you desire. Several orders have already come in -- supplies are limited. Because demand for pure ground beef (made from our cull cows) is so high, last fall we purchased two calves to help us meet this market. Although they were not born here, they've been here long enough that we think they are acceptable for ground beef consumption. We want to be totally open about our products and that's why we're telling you this.

We raised a couple of pigs for our household this winter and to our astonishment found that some of you love pork. The pigs tilled the garden for us and had a ball. The meat is very lean but firm and tasty because of all the rooting and plant material.

We've decided to offer pork for the first time. We will sell it by the half, hanging weight. We will cure the bacon, shoulders and hams, if you desire, using Teresa's old family recipe (sugar cure). All uncured meat would be available in mid-November. The curing takes about 6 weeks. If you want it smoked, you'll need to take it elsewhere to get that done. We're trying to stay simple initially -- maybe we can get more sophisticated in the future.

We also plan to add eggs by fall. The hens free range from a portable house that we move around with the cows. Vegetables are available on a surplus basis and in season. Ask if you want some. We hope to sell eggs and vegetables on broiler pickup days at least, and other days as available or desired.

We sincerely appreciate your encouragement and patronage. We're looking forward to serving you what we believe is the best meat in the world. Enclosed are a couple of business cards for you to pass along. If you have any questions at all, call. Come and see us.

Yours,

Joel F. Salatin, president

Polyface, Inc.

SPRING 1987 ORDER FORM

June BEEF

All prices based on per pound, hanging weight, and include cut, wrap and freeze.

____ Front quarter ($1.28)

____ Hind quarter ($1.60)

____ Half ($1.38)

____ Quarter ground ($1.19)

____ Liver ($1.60)

____ Heart ($1.30)

____ Tongue ($1.30)

November PORK

____ Half ($1.65)

CHICKEN

Indicate how many. Price is $1.00 per pound dressed.

____ July ____ October

____ August ____ November

____ September

Would you like eggs

($1.00 doz.)?

Yes No

Vegies? Yes No

If yes, what kind?

CLARIFICATIONS OR COMMENTS WELCOME (USE BACK IF NECESSARY). THANK YOU!

Spring 1991

Dear folks,

This is our annual formal communication and we trust that, although it is a little longer than normal, you will take just five minutes to read it thoroughly. We will bare our hearts to you a little, "sound off" a little, explain a little and help you a little. First let us sincerely thank you for your patronage and encouragement. We brag on you everywhere we go.

First, let's discuss transportation. As you know, we do not deliver. We want to encourage carpooling for those of you who want to spread the pickup burden. If you will enclose a stamped, self addressed envelope with your order form, we will send you names, addresses and telephone numbers of other customers in your general area. We realize some of you enjoy the drive out, looking at the animals and hiking around and do not want to carpool. That is fine. Don't send an SASE and we won't divulge the fact that you're a customer. Send an SASE, and we'll tell all.

Along that line, let us bare our hearts to you for a minute. The Polyface ministry is to produce the best food in the world. That is our focus. That consumes our energy, our time, our dreams and our expertise. Whenever we deviate from that focus, we begin to fail.

That failure shows up in a stressed lifestyle, economic setbacks, and food that may not quite measure up to our standard. We do not believe we are responsible to feed the world, or to build an empire. We enjoy the efficiency of smallness. In the world of living things, bigness has many drawbacks. Cleanliness, lack of stress, timely management and proper care are much harder to maintain on a large production scale. Tragically, many good small businesses become bad big businesses.

We do not intend to hire employees, and begin mass marketing. Nothing is as efficient as a small, diversified, wholistic family farm. We cannot hire the same level of care, hard work, commitment and expertise that we have as a family farm. While we may take apprentices or pass through offers, we do not want to become dependent on hired labor.

In the modern American business community, people get used to buying whatever service they want. Usually enough dollars, held out as the proverbial carrot on the end of the stick, can get movement out of anyone for anything. If a consumer wants a special consideration, a special service, he just offers to pay extra and it is done. But there is more to life than money, and Polyface will not be bought.

What we are driving at here is the whole issue of amenities and size. Bagging, delivering, freezing. These things seem small and simple, but multiply that by more than 350 customers from all over Virginia and out of the state, and then realize that when we process chickens we begin before 5 a.m. and have regular chores to do on top of the processing, and there simply are not enough hours in the day and enough energy in the individuals to do everything anyone may want ... at any price.

Now that is not to say that occasionally a car breaks down or there is an emergency that demands that we pick up the slack for you. All of you know that we have been willing to do that. But we can only take up so much slack. And offering to "pay for it" doesn't make it any easier for us.

In the past, we have generally assisted when a customer wanted his chickens bagged. This year we will have an extra one or two tubs and another table where those who want to bag can work while we go ahead and help the next customer. Sometimes our biggest need of the moment is to sit for a couple of minutes. Please do not think we are lazy. We may need a little breather so we'll have enough energy to take care of the chickens out in the field or fix supper. Our point here is not to complain, but to express our limitations. If we fail to understand our limits, we get cranky because we feel overworked and underpaid. Then the quality of our food becomes compromised because we can't get around to all the things that need to be done exactly when they need to be done.

Some people refuse to even think about driving out to pick up chickens. Yet they will spend hundreds of dollars on a vacation fling for personal enjoyment and eat food that's not fit to eat. We need customers who are willing to meet us halfway. We will gladly produce the world's best food. But the only way we can enjoy producing it at a reasonable price is to stop at the limits of our energy and efficiency.

Why does food of inferior quality cost twice as much as ours? Often the answer is amenities. Those refrigerated trucks and teamsters don't come cheap.

Most of you have and continue to be wonderfully supportive about meeting us halfway. You are wonderful. And our partnership will ensure a long and lasting mutually beneficial relationship. To people who want their food bagged, frozen, cooked and delivered, we only offer our apologies. This is a two way street.

Now to a related topic. Price. It has been said that the cheapest food is not always the least expensive. Certainly that is true. Conventionally produced food generally erodes the soil, pollutes the water, air and environment generally, makes people sick, hurts the rural economy and debases the food supply. What does all that cost? Nobody knows, but it would sure splash red ink all over 49 cent fryers and 79 cent hamburger. The point is that society, including future generations, picks up the tab for today's shortcuts.

People who are used to patronizing health food stores think our prices are extremely low. Others, used to more conventional food sources, think they run high. Regardless, ours is a true price and reflects everything. No hidden societal costs. Our prices are set to pay expenses plus a modest salary. The laborer is worthy of his hire. And the stereotypical "poor, dumb farmer" image is not one we endorse. Why should we not earn the equivalent of our customers' median income? We must receive enough compensation to keep us optimistic and joyous in our efforts. Our prices should not be compared with the local supermarket's. The philosophies and food qualities are as different as apples and oranges. While we promise not to gouge or to jack up the price because it's "natural," we also promise not to be agrarian serfs.

Thank you for bearing with us through this discourse. We've always wanted this newsletter to be more than commodities and business, and trust that this discussion has been helpful. Now let's move to business.

EGGS The eggmobile is in full production and we have plenty of eggs. Free ranged from a portable henhouse, these are happy birds. The eggs will keep up to 3 months in refrigeration because of their potency. Come on out and get 20 dozen.

BEEF This is our building year. We are keeping all of last year's heifers in order to increase the herd. That means we must temporarily reduce the number we sell, and that means we will be quite short this year. We do not know exactly how many we will have, especially the cows that we turn into ground beef. We tentatively plan to have two or three in June and about four to six in the fall. Please order now for the year and we'll assign what we have first come, first served, to previous customers only. We apologize for the delay. Eat more poultry.

CHICKEN The order blank dates are tentative. Last year the hatchery missed one week which threw two of the batches a week off schedule. We normally process Wednesday, Friday and Saturday of that week, which starts on Monday.

STEWING HENS This year we will have some stewing hens from the eggmobile. The old standby for these birds is chicken and dumplings. We like to cook several together and pick off the meat, then freeze it for quick addition to casseroles or chicken pot pie. The meat is more rich in flavor, but tougher than the young birds.

RABBIT Daniel, the enterprising 9 year old around here, is still desperately trying to fill all the rabbit orders that came in last year. We greatly underestimated the number of people who were interested in this most dense of protein meats. He also learned that his price was too low. Those who ordered last year and have not yet received rabbit will get it at the old price. New orders will be honored as rabbits become available.

VEGETABLES We plan to have cabbage, carrots, beets, tomatoes, yellow squash, zucchini squash, butternut and cushaw (pumpkin) squash, peppers, green beans, cucumbers, sweet potatoes and Swiss chard as available.

READY-TO-LAY PULLETS Available in early winter, these traditional American breeds come complete with range experience and a whole beak. Perfect for backyard home production, these pullets begin laying in January and last up to three years.

We were delighted with your support of Jon and Susan Moreshead's turkeys and lamb last year as they dovetailed their pioneer efforts with ours. Please turn this page over and catch up on their thoughts. Then fill out the appropriate order forms, enclose them both in the same self addressed envelope, put on a stamp, and enclose an SASE if you want carpooling. We'll sort it all out when it gets here.

Remember that we invite your visits, questions and criticisms. We are as close as the telephone. We sincerely appreciate your patronage and your loyalty and just can't wait to see you during the year to strengthen our relationship, building bridges between the rural and the urban, between producers and consumers, and between families and their future. Thank you for letting us serve you.

Warmly,

The Salatins of Polyface Inc.

Spring 1992

Dear folks,

Welcome to the new production year. It has been a year of unprecedented changes here at Polyface Inc. and we appreciate your being a part of this dynamic enterprise. Since this is our only formal annual communication with you, we'll use this forum to update you on our goings-on.

The first major difference you may notice when you come to the farm next time is a timber cut on the mountain behind the house. As many of you know, our farm extends up and over Little North Mountain, but we've never had access to that 400 acres of forest. Since coming to the farm in 1961, our family has always dreamed of a road "to the top of the mountain." After many years of searching and consulting, we traded about 50 acres of timber in four areas for 3.5 miles of all weather road. We can now harvest mature timber, clean up dead, down and diseased wood, and husband it. It opens up completely new and exciting stewardship possibilities. The cut areas will grow back quickly and the view ... well, you'll just have to see it.

The second major change is a 720 sq. ft. "RAKEN" house on the end of the equipment shed toward the house. What is a RAKEN house? It's a combination rabbit and chicken house. Daniel needed a permanent facility for his rabbit production and we wanted to research loose-housed, deep bedding laying hen production. We combined the two in a symbiotic relationship. The rabbit pens hang at eye level and the chickens are on the floor. The rabbits benefit from the warmth of the chickens' body heat in the winter, as well as the composting deep litter on the floor, and the chickens pick up bits of grain the rabbits drop and keep the bedding stirred, aerated and clean under the rabbit pens. It is perhaps the most soothing place on the farm to just go in and sit. You'll want to step inside when you come.

The third major change is that we just purchased ... uh, well, we hope you'll help us purchase, a brand new four-wheel drive tractor with a front end loader. It finally arrived a few days ago and we are excited about preserving Joel's back now. We've simply outgrown our ability to shovel everything.

Finally, we are nearly finished with the L shed on the far end of the equipment shed, and plan to use it for storing equipment and dimension lumber. The two-story greenhouse on the south end of the house is now painted and ready for the glass, which we now have in the poultry processing shed. We hope to get it put in before the first batch of chickens.

Isn't all this exciting? And how grateful we are to each of you for playing an integral part in our dream reality. How thankful we are that we have not gone the way of many in the organic foods movement; the way of conventional marketing methods where the producer and consumer do not know each other. Where California producers supply Virginia consumers. How much more enjoyable to build personal relationship bridges, to effect mutual appreciation between rural and urban, between producer and consumer. We trust that you will use our alternative food and our alternative thinking to pursue alternative choices in your lifestyle, entertainment, education and spiritual pursuits. We are in far more than the commodity business.

Last year's letter, you may recall, addressed carpooling. The response was unbelievable; NOBODY was interested. Well, a couple were, but that's all. We don't intend to address the issue again. It's dead. Thanks for coming.

POULTRY

On the order blank, we have articulated our tentative processing dates. We normally process Tuesday, Wednesday, Friday and Saturday. Remember that we call you a week in advance to get the exact day you'd like to get your chickens, but we thought it might be helpful to put it all down in black and white. Remember that we process in the morning and you come between 1 p.m. and 5 p.m. on that day. The best way to get the birds is in a cooler. You need about 2 quarts of cooler space per bird. If it is a warm day, we suggest you bring some ice. Poultry is extremely perishable. We buy bags in 1,000-count lots, so you can get them cheap from us if you'd like. You are free to bag them here; we provide a separate table for you to do that at your leisure.

The broilers, of course, are 8 weeks old, range raised in floorless pens which we move daily to fresh pasture. Our ration consists of corn, peanut meal, soybean meal, roasted soybeans, meat and bone meal, fish meal, alfalfa meal, kelp meal (seaweed), brewers' yeast and probiotics (*Lactobacillus acidophilus*). We use no antibiotics, hormones, coccidiostats, synthetic vitamins, germicides and the like.

Stewing hens are from our egg layers who have "served their time." All the comments about these last year were real positive. The meat's taste is very rich, but of course it must be cooked slow and long to be tender. These will be available late in the summer, as production begins to wane.

Ready-to-lay pullets, as usual, will be available around Christmas if you'd like the joy of having a backyard flock for your own eggs. We do not debeak these birds, of course, and raised on range they are good, aggressive foragers.

EGGS

As usual, we will have eggs from our "Eggmobile," the portable henhouse that sanitizes the pasture paddocks following cattle grazing. But we will also have eggs from our "RAKEN" house flock. Remember that we have eggs now, so go ahead and come out for them. Don't wait until May. Spring is the heavy production time. In order to have enough eggs for mid-summer, we always have an oversupply early in the spring. Apparently that's when God intended for us to eat more eggs, so let's join the seasonal roller coaster. Please save us egg cartons.

Be assured that these eggs are worth the effort, both for you and for us.

BEEF

A big apology is in order for all of you who ordered beef last year but did not get it. We had orders for twice as much as we had. We warned you, in last year's letter, that we were afraid that would happen. But, the good news is that the shortfall year is behind us and the herd is growing fast. This fall we should have about half again as many as we did last year. We may still be a little short, but we hope it won't be as bad. Next year we should be in good shape. Our beef is grass fattened, raised on perennial polycultures in an intensive controlled grazing program that seeks to mimic the two great herbivore principals: herding (density) and movement (duration). You may notice that we are not offering June beef any more. We believe the fall beef tastes better, having just come off lush fall pasture, and harvesting at that time capitalizes on natural forage and weather cycles. The only beef we may have in the spring is from a cow that loses her calf. In that case, we make ground beef and try to move that to those who want ground beef only. You may note on the order blank if you would prefer to have it in the spring. Remember that the spring, though, is an indefinite deal.

RABBIT

Daniel, at 10 years old, is our resident rabbit expert. Again, our apologies for being so far behind in filling the orders. Two years ago when he had two rabbits and got orders for 150, he thought he'd never catch up. But you know rabbits. He is not feeding them the way conventional wisdom suggests: pellets only. Rather he is feeding them free choice hay, oats and pellets. It's surprising to see what a difference the forage makes in the fat on the rabbit. Of course, rabbit is the only guaranteed no cholesterol meat because its fat stays outside the meat under the hide or inside in the organs. Having gone through the learning curve, he is expanding rapidly, improving the genetics, and produces a fine rabbit, as many of you can attest. Because rabbits produce year-round, he is having trouble matching orders to chicken orders to save folks an extra trip. We have decided to fill orders from you who live far away during the summer when you are coming for chicken. You who live nearby, we hope it's not too much to ask you to come off-season, like fall, winter and early spring, to get your rabbit. It might help out on freezer space too. If all goes well, he should catch up on orders by next spring.

VEGETABLES

Remember that these are available when we have surplus.

FIREWOOD

For heating or romance, wood fires are hard to beat. It's regenerative and doesn't produce the harmful particulates of other fuels. We cut it and pile it here at the house for you to haul at $25 per pickup load. Call us if you want it delivered.

Finally, we're hosting another Field Day this summer, July 11. This is your opportunity to mingle with folks from around the country, see inside our farm and get the scoop on what we're trying to do. A hay-wagon tour with barbecued Polyface chicken and trimmings will add to the day's festivities. Reservations must be made and you Polyface patrons may come at half price ($10 per person, $17.50 per couple, students $5 and children under 12, free).

Please, if you have any questions, call. We now have an answering gizmo. Feel free to bring a picnic and see how we do things. We learn from each other. Thanks again for your faith in us -- we don't take it lightly. Remember the orders are first come, first served, so don't dillydally with this thing. Thanks.

THE SALATINS OF POLYFACE INC.

Spring 1993

Dear folks,

"I just can't eat that stuff out of the store anymore." We never tire of hearing you say that. We appreciate the seemingly hundreds of times we've heard that statement, and trust we will never violate your trust in producing the world's best poultry, beef and rabbit.

Several of you have called, lamenting the fact that your freezers are well nigh empty and May is still many meals away. We're sorry, but you'll just have to order more birds this year. Cycling our production with nature's seasons must not be compromised. As soon as we produce food against the seasons, disease, costs and environmental degradation escalates. We calve when the deer are fawning, start chicks when wild turkey eggs are hatching, and increase the work load when the days are longer.

We've enjoyed the winter respite, and have been busy spreading our message to other folks around the country. In November, Joel keynoted the Carolina Farm Stewardship Conference in Rock Hill, SC. In early January, he keynoted the Maryland Organic Food and Farming Association annual meeting in Annapolis. The end of January and early February, he spoke at the Texas Eco-Fair in Austin, the Northeastern Grazing Conference in Lancaster, PA and keynoted the West Virginia Direct Marketing Association annual conference in Morgantown. On three of these the family went along, enjoying a mini-vacation.

Many of you know that we have written a 60-page *Pastured Poultry Manual*, and it has sold nearly 1,000 copies in less than 2 years. It has stimulated probably 100 folks around the world to begin raising chickens and eggs using the Polyface model. This winter, we have updated it, increased the information by 50 percent, and added 20 pages of pictures. Lord willing, it will be released as an honest-to-goodness 200 page paperback book sometime in May or June. We've been told that we're starting a revolution in agriculture. Isn't that exciting! You folks deserve much of the shared credit, for standing with us and turning our dreams into reality. Thank you.

We hope this year to put together a brochure about Polyface, and introduction to our methods and philosophy, so that newcomers can quickly understand the game plan and you old-timers don't need to be bored in these letters with unnecessary introductory information every spring. We are taking one step at a time.

RABBIT

Your hunger for Daniel's (the 11-year-old Salatin) alternative rabbit has far outpaced even the ability of rabbits to keep up. He has meticulously recorded all your orders, and knows how far behind he is. For this year, we are not taking orders, but rather focusing all attention on catching up with old orders. His goal is to catch up by year's end, double the breeding stock, and be ready to accommodate new orders next year. Thank you for your patience. Please feel free to take a look at them when you come out.

EGGS AND READY-TO-LAY PULLETS

For the first time it appears we will not be able to keep up with eggs this summer. We had planned to keep at least 100 pullets for layers this year (out of the 350 we started last summer) but the demand for these pullets rose sharply last year and we finally began turning people away when we only had 50 left. As a result, our layer flock will be under 200 this year. We'll squeeze those old gals all we can. Remember there is nothing like having a dozen or so in your backyard to produce your own eggs. We use nonhybrid old American varieties, and encourage you to get some to produce your cackleberries.

CHICKEN

Is it really any different? You bet. We've articulated it this time as an educational tool -- and to help you understand why it's worth a premium price. At the end of this comparison, we'll pick up this letter and close with BEEF. *[See "What is the Difference," beginning on page 309]*.

Quite a list, huh? It is amazing how far off base a production/ processing model built on erroneous beliefs can deviate from what is good and reasonable. If you have any questions about it, please ask. We tried to cram all the information we could on that one sheet of paper.

We always hate to see prices rise without any explanation. You will notice that the chicken price is up a dime a pound, or 8 percent. This is the first price increase in 3 years, and during that time the cost of our chicks has gone up a dime; taxes and insurance have risen dramatically; fuel and utilities (electricity) has increased and our feed bill is higher. We feel badly about having to raise the price, especially with Clinton's pledges to increase taxes, but we must make a profit. Thanks for understanding.

Our last two batches last year were on the small side, and we believe the eggs came from an old flock just going out of production. We are researching alternative chick sources, and if we must change chick suppliers, this will greatly increase our cost of the chicks. For sure, we want those 4-pounders every bit as much as you do. At least you are paying by the pound, so smaller birds are cheaper. You just have to eat more of them. Again, thanks for bearing with us.

BEEF

Again we must apologize for not being able to accommodate all the orders last year. We are expanding as fast as we can. We're coming closer each year to having enough to go around. Everyone seemed pleased with the changes last year at the slaughterhouse, and our meeting you there. We plan to continue that policy. It made Mullins happy, too. Remember that the size differences from beef to beef is quite dramatic since we do not only steers, but heifers also. If you'd like a small quarter, indicate that on the order blank. Do the same if you'd prefer a large one.

You may notice that the price on front quarters has stayed the same, but the other prices have risen slightly. With this huge price spread, the best value for your money is the front quarter. It seems a hopeless cause to spread the price enough to keep fronts and hinds in equilibrium. If we could breed a cow with all hind quarters, we'd really have something — kind of like a pig that's all ham. But we can't (despite what the biotech monkeys think) or shouldn't, so we're stuck with fronts and hinds. Now please, don't abandon the hinds just because we're making a big to-do about this. We're just trying to explain the price difference and level with you about our needs.

That just about wraps it up for this spring. Fill in the order blank, put it in the enclosed envelope, stick a stamp on, and fire it back to us. Remember it's first come, first served. When we reach our capacity, that's it. We're sure looking forward to seeing all of you again this year. While we do operate a business, we prefer to think of it as a relationship-builder; and while you are customers, we prefer to call you friends. You're welcome anytime. Come for a hike, to work off some flab, to enjoy a picnic or whatever. You are collectively the greatest boss we could want. Fill out the order blank and have a great season.

THE SALATINS OF POLYFACE INC.

WHAT IS THE DIFFERENCE?

POLYFACE CHICKEN	CONVENTIONAL CHICKEN
Unvaccinated	Vaccinated (immuno-suppressant)
Full beak (no cannibalism)	Debeaked (cannibalism a problem)
Probiotics (immuno-stimulant)	Antibiotics (immuno-depressant)
Composting litter in brooder (sanitized through decomposition)	Sterilized litter (sanitized through toxic fumigants and sprays)
Carbon/Nitrogen ratio 30:1	C/N ratio 12:1
Practically no ammonia vapor (smell)	*Hyper-ammonia toxicity
Brooder skylights	*No skylights
Rest at night -- lights off	Artificial lighting 24 hours/day
No medications	Routine medications
No synthetic vitamins	Routine synthetic vitamins
No hormones	Routine hormones
No appetite stimulants	Routine appetite stimulants (arsenic)
Natural trace minerals (kelp)	Manufactured and acidulated trace minerals

WHAT IS THE DIFFERENCE?

POLYFACE CHICKEN	CONVENTIONAL CHICKEN
Small groups (300 or fewer)	*Huge groups (10,000 or more)
Low stress (group divisions)	*High stress
Clean air	*Air hazy with fecal particulate (damages respiratory tract and pulls vitamins out of body, overloading liver)
Fresh air and sunshine	*Limited air and practically no sunshine
Plenty of exercise	*Limited exercise
Fresh daily salad bar	*No green material or bugs
Short transport to processing	*Long transport to processing
Killed by slitting throat (per Biblical directives - see Leviticus)	*Killed by electric shock (inhibits bleeding after throat is slit)
Carefully hand eviscerated	Mechanically eviscerated (prone to breaking intestines and spilling feces over carcass)
Processing uses only 2.5 gal. water/bird	Processing uses 5 gal. water/bird

WHAT IS THE DIFFERENCE?

POLYFACE CHICKEN	CONVENTIONAL CHICKEN
Guts and feathers composted and used for fertilizer	Guts cooked and rendered, then fed back to chickens
Effluent used for irrigation	*Effluent treated as sewage
Customer inspected	*Government inspected
No injections during processing	Routine injections (anything from tenderizers to dyes)
Low percentage rejected livers or carcasses	High percentage liver rejects or carcasses (breast blisters)
Dead birds fed to buzzards or composted	Dead birds incinerated or buried (possible contamination of water)
Sick birds put in hospital pen for second chance -- most get well	Sick birds destroyed
Manure falls directly on growing forage and active soil for efficient nutrient cycling -- converted to plants	Manure fed to cattle or spread inappropriately (ammonia vaporization --air pollution; nitrate leaching--water pollution)
Fresh air and sunshine sanitize processing area	*Toxic germicides to sanitize processing facility
Cooking loss 9% of carcass weight	Cooking loss 20% of carcass weight

WHAT IS THE DIFFERENCE?

POLYFACE CHICKEN	CONVENTIONAL
Long keepers (freeze more than a year)	Short keepers (freeze only 6 mos. or less)
No drug-resistant diseases	Drug-resistant diseases (R-factor *Salmonella*)
Low saturated fat	High saturated fat
No chlorine baths	Up to 40 chlorine baths (to kill contaminants)
No irradiation	FDA-approved irradiation (label not required)
Environmentally responsible	Environmentally irresponsible (hidden costs)
Promotes family farming	Promotes feudal/serf agriculture
Decentralized food system	Centralized food system
Promotes entrepreneurial spirit	Promotes low wage/time-clock employment
Rural revitalization	Urban expansion
Consumer/producer relationship	Consumer/producer alienation
Rich, delicious taste	Poor, flat taste
Edible	Inedible

Also applies to nearly all "certified organic" chicken.

Spring 1994

Dear folks,

Are you sitting down? . . . Go ahead. Now, that's better. It's been the best of years and the worst of years. You and us are getting along great. How do you like being told by the government that you're not capable of deciding where to buy your food? We'll get into that in a minute, but first, some of the good news.

1993 was a banner year in many ways. We produced and processed 8,000 broilers, 24 beeves, 2 pigs, 3,000 dozen eggs, 250 rabbits, 250 pickup loads of firewood and a pile of logs for lumber. And it's not without notice.

This winter, Joel has spoken in agriculture conferences in Mississippi, Missouri, Iowa, Minnesota, Ontario, Wisconsin, Pennsylvania and Ohio. Whew! What a winter. In December we released ***Pastured Poultry Profits***, a 340-page book detailing our philosophy, production, processing and marketing of broilers and eggs. A companion project, a 45-minute professional VHS video by the same title, offers a fast-moving visual/sound synopsis of the broiler model. Both have been selling extremely well and are getting excellent reviews in the alternative agriculture press. The book sells for $30, video for $50 and package for $75. Around the U.S., at least 200 pastured poultry enterprises are up and running, mimicking the Polyface model. What an exciting time to be in agriculture. We thank each of you for making all this possible -- give yourself a pat on the back.

But now the bad news. Wouldn't you know that just when we were making converts around the country, the feds would come in and try to shut us down? The enclosed copy of a guest editorial I wrote for ***The New Farm*** gives you the basic pitch, but let me flesh out a little of it for you. First, the inspectors have not said a word to the many other farmers who are marketing exactly the way we did through Mullins (they estimate at least 50). In other words, we were clearly singled out for action. Second, all the way through the chain of command, the bureaucrats blamed supermarkets and big meat and poultry packers for their aggression. They don't want any competition, and they are unwilling to admit that it is much easier to keep a small operation cleaner than a large one.

Third, although our poultry is alright for now, they told us unequivocally that if more people begin raising and processing chickens like we do, that they will try to shut down our chickens also. Fourth, their objective is to protect YOU from unsafe meat and poultry, and YOU cannot be trusted to decide what is right and wrong. After all, the large processors have YOUR interests at heart; we're just hayseeds out to make a fast buck and capitalize on paranoia. Right?

We have spent many hours and dollars all winter with politicians, state officials, federal officials, our attorney, and are not getting very far. We are not trying to pass on any of these costs to you -- but we are going to ask for your help, in a minute. Let me just get something off my chest first. You know, it's easy to mail a check to Greenpeace, Sierra Club or the like, and assume that they'll "take care of it," whatever "it" is. Did you know that in recent years the environmental groups have wanted big, corporate farms because they are easier to control (fewer of them) with environmental regulations? Do you really think your food is safer coming from a corporate conglomerate or from your community?

How do we respond when **60 Minutes** shows the filth in beef and poultry plants? Often it's: "regulate those things. Put the screws on them! They shouldn't be allowed to do that!" Please, please, look at the other side. The regulations are size neutral, and have nothing to do with quality, but system. For example, in order for us to have an inspected poultry processing facility, we need a bathroom. Why? For our employees. We have none. "Well, for you then," the inspector said.

"But why? We have four bathrooms in two houses 50 feet from our processing area. And the men can walk around the other side of the tractor and go, for that matter," I responded.

"That's what the regulation says," he quipped. End of discussion. Now folks, how important is it that we have a bathroom (with approved septic, approved toilet, properly isolated from the processing area, etc., etc.) in the processing facility? See what happens when we demand salvation by legislation? It just shoves the little guys out of business, and denies YOU the freedom to choose alternatives. A society cannot be absolved of the decision-making risk without also losing its freedoms. Risk and freedom go hand-in-hand.

It's the same issue as denying homeschooling, denying alternative medical care, denying alternative retirement programs (instead of Social Security) and a host of other things that plague our once-free country. Our 7-year-old daughter Rachel wants to make pound cakes for you folks this summer -- and they are good (she's been practicing). But without an "approved" kitchen, with all sorts of expensive gingerbread, it's illegal. The pork we sold last summer was illegal. Listen, folks, the people dying and getting ill from tainted food aren't getting it from small, local farmers -- they're getting it from inspected food empires. In Virginia, we can have abortions but not buy a glass of raw milk from a farmer. We can buy handguns but not a pound of sausage from a farmer who puts a pig on a pole at Thanksgiving. Believe me, the environmentalist groups are not lobbying for decentralized agriculture and food choice. But the most efficient way to change the food empire/cartel is competition from small, independent, high quality local producers meeting the demand of like-minded folks who want to opt out of conventional food channels. We've practiced using the stick for change too long; it's time to use the carrot. Instead of "there ought to be a law," we should say:

"there ought to be some liberty."

Well, where are we? Our Del. Vance Wilkins has submitted a LIVE ANIMAL PURCHASE CONTRACT to the Virginia Dept. of Agriculture and Consumer Services, and they turned it down. He has contacted our Congressman, U.S. Rep. Bob Goodlatte, to see if the federal folks will okay it (the state officials are scared of the feds). If they turn it down, we will have to truck the beeves and pigs an extra 30 miles to T&E Meats in Harrisonburg (federal inspected) for slaughter, and they will use their refrigerated truck to bring the carcasses back to Mullins for processing. Mullins will lose hundreds of dollars of business as a result. As long as the state doesn't come after the rest of their clients, they can survive. But if this happens to a few others, Mullins will have to close. Aren't these bureaucrats true public servants? It will add cost, stress the animals, take far more of our time, and be a logistical nightmare. It makes my blood boil -- and it should yours.

Here's what we'd like you to do: write a letter or make a phone call THIS WEEK telling Bob Goodlatte that you think it is outrageous that the USDA is trying to put small farmers and community custom processors out of business by re-interpreting federal regulations to assume that regardless of contracts or agreements, a hanging weight price cannot be the basis of a live animal purchase. Mention us by name -- he needs to get the picture. Tell him you want to be able to buy locally-produced meat, milk and poultry and you don't like the government being the Gestapo of the huge packers and processors. Enough said. Thank you for ACTING.

U.S. Rep. Bob Goodlatte
214 Cannon Office Bldg.
Washington, D.C. 20515
(202) 225-5431

Obviously if you want to contact other federal delegates or senators, that's fine too. The bottom line is this: people like you and us threaten the very foundations of conventional thinking, and they will try to do anything possible to keep you shackled to approved channels, and us from producing and marketing outside approved channels. We're in this together; let's not let our grandchildren down. Now on to better topics.

CHICKEN As usual, we have 7 batches throughout the season. This year we'll have a little label that goes on your cooler or bag with our inspection-exempt number on it (Va. #1001 -- the only one in Virginia). That allows us to go up to 20,000 birds as a producer-grower. Remember everything is first come, first served, and our capacity is limited. Typically, the last batch is the first to fill up. As the first and last batch fill, we'll begin shifting later orders toward the middle until they are all full. Another family, located in Burnsville, Bath County, is doing a fantastic job duplicating our model and we may use them to help us meet the demand. They are using our feed, buying chicks with us -- the whole 9 yards. You would find the Shell family like-minded in every way. Feel free to actually circle a day on the order form -- we'll just send a postcard reminder in that case. Otherwise, we'll call a week in advance to let you select one of the four days.

EGGS Please don't wait until summer to get eggs. We have plenty, plenty, plenty. They keep a long time (2 months) and offer dramatic differences to the supermarket fare. Through May at least we'll give a free dozen with every 10 dozen purchased.

STEWING HENS These are our older hens who have completed their second laying season. We don't have very many this year, so we're sure they won't go around. The taste is rich and meat needs to be cooked differently than broilers to be tender.

READY-TO-LAY PULLETS Again, we encourage you to produce your own eggs in your own backyard. The joy and satisfaction of producing some of our own food is truly wonderful. These are all heavy breed, nonhybrid, brown egg layers. We'll have Rhode Island Reds, Barred Rocks, White Rocks and probably at least one other breed.

RABBIT You will be happy to know that Daniel, the 12-year-old man around here, is expanding his rabbit enterprise and will be able to serve most of you this year. Remember that rabbit is the only cholesterol-free meat and contains the highest protein/pound of any meat, allowing small portions to be quite filling. Last summer we perfected a pasture grazing program for the young stock and they never did better. You'll want to take a look at them out in the field. Since the rabbits are available year round, we'd like to take care of you folks who live far away when you come for beef, pork or chickens, and get you close folks in the off-season. If you'd like rabbit before the first batch of chickens, indicate that on the order blank, because we have some fresh ones in the freezer right now.

PORK Many of you got to sample (illegally) our "Pigaerator Pork" last year. We used these two body-builders to aerate and make our compost in the hay shed after the cows went out to pasture in the spring. Two pigs turned and aerated 75 cubic yards of bedding in 8 weeks, revolutionizing compost economics. The grain I had put in the bedding as their incentive fermented in the anaerobic medium and produced the sweetest, most delicious pork and sausage we've ever eaten. We're sorry that it ran out so fast. This year we're putting in 6 and doing both bedding sections. We plan to process 3 after they finish their pigaerating and then put 3 out on pasture in what we'll call our "Tenderloin Taxi," a portable pen we'll move around like chicken pens. Once the pigs get fairly large, they can convert grass efficiently, and we'd like to see how they do on pasture with grain supplementation.

Because of our inspection hassles, we want to sell all of them a half at a time -- feel free to invite someone to "go in" with you. We aren't sure yet if we can sell it by the package even if we get it federally inspected -- we think we'd have to get our freezer and room inspected. It's ridiculous, I know. It's hard to get a straight answer from these officials -- the answer changes every Monday morning. But if you pick the wrong Monday morning, you're guilty until proven innocent with these Nazis. We've put the hanging weight on the order form to help you see what you'll pay for. Remember that processing drops the weight by 25-35 percent, so you won't actually bring that much meat home. Meat is heavy: high weight, low volume. You eat more than you think.

BEEF Major shake-up this year. For more than a decade we've been battling the front quarter problem. Nobody wants a front quarter -- it's about as lonesome as a hymn's third stanza. Mullins gave us some wise and wonderful counsel last fall -- SPLIT HALVES. This year we will not offer front and hinds, but rather split halves. This way you get some of the front and some of the hind, letting everyone get some of those prime hind cuts. We've also adjusted to price to give you a nice break for getting a half. Again, we've plugged some numbers in there so you can see your options. As usual, let us know if you want a smaller one or larger one -- there is a lot of latitude because we process both steers and heifers. If size doesn't matter, that's fine too.

The only down side to this change is that you won't have quite as much latitude in your processing options. You can't make steaks and roasts out of the same piece of meat. But we really believe it's going to be a big improvement over forcing so many of you to take nothing but fronts. Of course, you'll decide about packaging -- how many per/how big. Mullins says you can give them "number in the family" and they can fix you up fine. One word about price: remember that if you went to a meat counter and bought the pieces by the cut that would come out of a beef, you would be paying $2.20-

$2.30 per pound equivalent hanging weight. In other words, our prices, because you're buying in some volume, are much cheaper per pound than if you bought the equivalent meat by the cut a little bit at a time out of a meat counter. Ground quarters will be extremely limited this year. We're afraid again that we won't have enough beef to go around this year, but we'll do our best. We're expanding the cow herd as fast as we can, but there's a long lag time on these gals. Because you enjoyed our poultry comparison chart in last year's newsletter so much, we did a similar one for beef. It's amazing how far wrong something can be, isn't it?

We apologize for the length of this newsletter. Be assured we've agonized over it for some time, but we knew we had to give you enough information to explain where we are and you are, and we hope we've done that. Please, if you have ANY questions or comments, call or write -- communicate. Remember that everything is first come, first served, so don't dilly-dally with this order blank. Fill it out and stuff it in the enclosed envelope, slap on a stamp and fire it back. For you folks that are new, we assure you that this is a very unusual newsletter -- don't be scared off. We pledge ourselves to produce the finest, purest, most nutritious food in the world -- thanks for appreciating it. We can't wait to see all of you this season. Many thanks, and blessings on you all.

THE SALATINS OF POLYFACE, INC.

MORESHEAD TURKEYS

Although we had a disappointing year in terms of high mortality and consequent problems meeting orders, we learned a lot. These problems, however, have opened the door to some pretty exciting developments. It will take more space to tell about them than we have available in the Polyface newsletter so look for a separate communication to be sent soon. If you have ordered turkey in the last couple of years you are on our mailing list and will get our information and order form; if you haven't ordered turkeys for a couple of years be sure to call with your name and address so we can send you your copy. Our phone number is: 703-885-2304. Bon appetit!

HOMEPLACE FARMS
ALL NATURAL LAMB

Homeplace Farms is a small family farm located in Shenandoah County, Virginia. We are proud to offer the customers of Polyface our all natural lamb! Joel Salatin has been gracious enough to allow us to include this info-sheet in this mailing so that those of you who have expressed interest over the years would become aware that we are available and be able to order at the same time you are ordering your Polyface products. Please just include this form when you return your Polyface order.

Our lamb is high quality, tender, young lamb of no more than six (6) months of age. Mother's milk and high quality forage has been its only diet. It will be available for pick-up in October (some special orders for earlier lamb are possible). Approximate Hanging Weights run from 40 to 60 pounds per lamb. God Bless you and thank you for your interest.

-Mark & Laura Accettullo (3/94)

General Information:

1. As a small farm and in order to keep our prices down we can only offer bulk orders of meat.

2. All prices are based on "Hanging Weight" per pound and includes "cut, wrap and freeze" to your specifications. Hanging Weight is the weight of the carcass just before it is carved into your specific cuts. Therefore, there will be a difference between the Hanging Weight and the total weight of the meat you take home. This is due to bones and other carving waste that are discarded.

3. Our meat is butchered at a locally owned, state inspected shop of the highest quality.

4. After your order is placed, we will follow up with you on your desired cuts, etc.

5. All orders must be paid for upon pickup.

ORDER FORM

Qty.	Option	Price/lb.
A._________	WHOLE	2.85
B._________	HALF	3.25

NAME: ______________________________

ADDRESS: ____________________________________

____

PHONE: ______________________________

Homeplace Farms, Box 472, Woodstock, VA (703) 933-6916

DIRECT MARKETING: LET'S KEEP IT ALIVE

by Joel Salatin (article from *New Farm* magazine, 2/94)

"You did what?" I yelped incredulously at my unexpected visitor.

We've impounded your beef at the slaughterhouse, repeated the man at the front door. He had just flashed his big bronze badge: Compliance Officer, Food Safety and Inspection Service, USDA.

"But why?" I asked We were following the same procedure we'd always used.

We had a complaint that you're selling uninspected meat.

"But we sold that meat by half or by quarter six months ago, while the animals were walking around in the field. Buyers' names are on the carcasses now," I said, feeling like a child sent to the principal's office for something I didn't do.

Our informant told us your price is based on carcass weight. No one can touch or move that meat until we complete our investigation, the officer responded coolly. It might take six months. And if your price is based on carcass weight, we can condemn it all. It will be sent to the renderer.

"But in six months, it will spoil. This is our livelihood. We're talking $10,000 worth of beef here. By the way, is anything wrong with it?" I asked.

No, that has never been indicated. You just can't sell uninspected meat. He added that he could not be responsible for all the meat being lost.

He left. "This is my tax dollars at work?" I fumed.

For 30 years, our family has produced and marketed organic, grass-fattened beef. More than 400 customers depend on our beef, pastured poultry, pork, range eggs and forage-based rabbit.

The compliance officer showed up on our doorstep Sept. 21, a day after we had taken 14 forage-finished cattle to our local custom slaughterhouse. For several sleepless days and night, we dealt with bureaucrats who changed their minds every few hours.

Thanks to our persistent efforts and some help from our U.S. representative, we were able to sell this batch. But our way of doing business is now against the rules, and our marketing for '94 is in limbo. Regulators say we can sell either by liveweight or by the head. Both options are unfair to farmers and to consumers because they fail to account for the wide variations in the dress-out percentage between animals. The difference could amount to $150 per animal.

New policies that suddenly put us at odds with the feds could affect anyone involved in small-scale processing. It's time farmers take the initiative to keep direct marketing alive.

All the meat inspectors I've talked with, from the state level to the federal, agree that this recent effort to complicate carcass-weight sales for small producers is being pushed by the retail industry. This sector wants to boost its market share at our expense. As Dr. Charles Grosse, director of meat inspection for the Virginia Department of Agriculture and Consumer Services lamented: "What am I supposed to tell the retailer who complains that you are selling meat without going through all the hoops he has to go through?"

Changing Custom

In Virginia and many other states, there are three levels of inspection: federal, state and "custom." Federal inspection requires an inspector to be in the plant and allows the meat to be transported across state lines and sold by the cut at a per-pound rate. State inspection is essentially the same thing, but the meat can't be shipped across state lines.

"Custom" allows slaughter and processing without inspection so long as the meat is sold prior to slaughter. Each package must be stamped "NOT FOR SALE." Custom facilities are generally small and serve local communities. Our county alone has half a dozen.

Historically, custom processors served local farmers who brought in animals and took home meat for their own use. But as rural areas became more urbanized and per-capita beef consumption declined, this clientele dropped off. When folks didn't want a whole beef, farmers began selling quarters and halves. As long as the arrangements were made while the animal was alive, the custom plant could legally process the meat.

When we take animals to the slaughterhouse, our instruction sheet clearly delineates each animal by ear tag and indicates who owns which part. Owners' names are later affixed to the carcass. Thousands of us add value to our meat products this way.

The new government position is that any agreement that sets a price based on carcass weight is an "after slaughter" sale. I say hogwash. My customers saw the animals in the field, they know exactly how they were raised, and most plan months ahead for the meat to arrive.

By contrast, the person walking into a retail store has no idea where or how the meat was raised, made no prior arrangements with the producer (or retailer), and makes a decision to buy long after the animal was slaughtered.

Looks like a different situation to me.

Although we farm organically, this is not an organic farming issue. It is an issue that spans small business, rural economic vitality, animal welfare, consumer choice and market access. It has nothing to do with food safety or promoting the general welfare, since custom-slaughtered meat doesn't reach the mainstream meat consumer.

Small is Better

There are many advantages to small processing houses that make them work fighting to preserve. I'll describe three:

- Product integrity is maximized. In direct marketing, product integrity is everything. Federal and state facilities, because of offices and bathrooms for inspectors and other costly overhead items, are inherently larger plants. They have to do more volume just to cover the costs of regulatory requirements. The larger the facility, the more difficult it is to maintain product integrity. I know from experience that bigger plants can't keep track of identified meat as well as my small one can.

- Stress on the animal is minimized. This bears directly on meat quality. The longer the animal is transported, and the longer the animal waits in the plant before slaughter, the more its muscles tighten. It is inhumane to transport animals long distances when they could be processed close to home.

- Cost is lower. A higher level of inspection pushes up processing costs, making meat more expensive. The more consistently we keep our food in alternative marketing and processing channels, the greater our ability to hold costs down and garner market share.

We in organic agriculture often shoot ourselves in the foot by producing alternatively but then processing and marketing conventionally. Big operators are most efficient at materials handling and transportation. They haul tractor-trailer loads and we haul pickup loads. If we try to compete on their terms, the price of alternative food goes through the roof.

The whole issue of who can sell what to whom spans far more than meat products. Many current food regulations are arbitrary. Either rules originally designed to protect consumers have snowballed to ridiculous extremes, or they have been twisted to strengthen the market position of corporate interests.

I can raise and process 1,000 chickens, and even transport them anywhere in the state, but I can't sell a gallon of milk without a multi-thousand-dollar facility. I can dress all the rabbits I want to in the backyard on a clothesline, taken them to town and sell them to restaurants, completely free of inspection. But if I butcher a hog out back and sell one package of sausage to a neighbor, I'm a criminal.

We have an odd mixture of choices in this country. I could if I wanted to apply pesticides by the gallon, but technically I can't compost a single dead chicken without a permit for an approved facility.

Is there rhyme or reason to these laws? What have we allowed misdirected regulations to do to our food system?

For farmers and consumers who want to continue a traditional direct-sale relationship, there ought to be specific low-volume amounts of meat and food products exempt from inspection. The current federal poultry producer exemption from inspection for up to 20,000 birds is a prime example--and an entrepreneurial gold mine for small-scale producers.

I'm not arguing to eliminate all meat-inspection regulations. But, really, if there is no need to inspect 20,000 chickens per year from a farm, where does the public welfare become endangered? At one prepared pot pie? At 1,000 gallons of milk?

I maintain the difference is one of scale, not of livestock type or commodity. It is time to recognize how much easier it is to keep things clean where we are dealing with small numbers. Taking chickens from the pasture every couple of days and processing them in our backyard open-air shelter is far more sanitary than continuously slaughtering confinement birds on an assembly line.

When customers are satisfied with the cleanliness of a farm and the integrity of the food produced, why should they be denied the right to purchase food there?

If sustainable agriculture tries to enter the marketplace at the mega-food-industry level, it may never become what we all wish it would. We certainly won't build relationships with individual clients or keep dollars in our local economies. Centralized food systems are inherently unsustainable for the land, unrewarding for farmers and inhumane for livestock, regardless of the production models used.

We need to develop distinctly sustainable marketing that maximizes decentralized processing and distribution. We also must create more value-added economic opportunities for individual farmers.

For starters, let's champion legislation something like this:

BE IT RESOLVED that all agricultural products processed and marketed on or within 20 miles of the producing farm shall be exempt from government inspection up to an including the following annual limits: 20,000 head of poultry (current regulation); 300 cattle; 500 hogs; 1,000 sheep, lamb or goats; 50,000 gallons of milk, or 50,000 pounds of cheese, yogurt, butter or ice cream.

The list could go on, but you get the idea. It is a very limited exemption that requires buyers to come to, or very near, the farm--close enough to inspect it. Small producers who direct market tend to be extremely clean because we have a limited, known clientele that keeps a close eye on us.

The current inspection exemption for poultry and rabbits works well. I'd argue that the propensity for unsafe food reaching consumers from farms under these expanded exemptions would be no greater than it is under current inspection rules for commercial slaughterhouses. Electronic communications, consumer awareness, stainless steel, refrigeration and rural electrification make this era distinctly different from the farm settings of 40 years ago--the ones regulators tell horror stories about.

We've become accustomed to horror stories of a different kind: news reports about contaminated processed food and meats, accelerated processing lines and slashed inspection budgets. Plants that receive diverse shipments of livestock and sent meat to distant customers may need inspection and regulation, because the risks of contamination are great, and few consumers get in to visit.

But there is clearly a place where inspection is inappropriate because it is unneeded. That situation includes small-scale farmers who sell to their neighbors and have livestock processed in small, local slaughterhouses. Inspection imposes industrial-strength controls on kitchen-sized--and kitchen-nice--facilities where no demonstrated problem exists.

Again, food safety is not the issue. Only conscientious entrepreneurs contributing wholesome products will survive in local markets where the products are directly linked to the people who grow, pack and sell them. At issue is whether farmers can sell food to people who trust them.

It is time for our laws to encourage ecologically sustainable farming, cconomically sustainable rural communities and a nutritionally sustainable food system.

Editor's Note: Joel and Teresa Salatin and family operate Polyface, Inc., their pasture-based farm in northwest Virginia. (See "Profit by Appointment Only," **The New Farm**, Sept./Oct. '91). Since he wrote this commentary, they've been forbidden to sell more beef in their usual manner. They will revise marketing procedures to continue using the same local slaughterhouse. To promote the small-scale food-processing exemptions he outlines here, Salatin is starting a nonprofit group called Consumers and Farmers for Choice *[note - this has now changed to Food Alternatives with Relationship Marketing, or FARM, and is still in its organizational stages]*. For details, send a self-addressed, stamped envelope to: FARM, P.O. Box 186, Willis, VA 24380.

FOUNDATION STATEMENT, CONSUMERS AND FARMERS FOR CHOICE (now FARM)

WHEREAS, small scale food production/processing entities deal in small volumes inherently easier to sanitize than large facilities with large volumes; and

WHEREAS, costly, unnecessary and inappropriate inspection requirements on small scale producer/processors unfairly restrict production/processing facilities; and

WHEREAS, unnecessary inspection requirements form an economic barrier to small scale producer/processors entering local, alternative market channels; and

WHEREAS 40 miles is close enough for any customer to personally inspect an establishment; and

WHEREAS value adding to agricultural commodities through direct marketing is essential for small farm survival and rural economic stimulation; and

WHEREAS stimulating market access for small scale producer/processors is essential to decentralize agriculture; and

WHEREAS a decentralized agriculture stimulates regional self-sufficiency, humane production/processing models and fairer market competition; and

WHEREAS some small scale production/processing commodities like poultry and rabbit currently enjoy inspection exemptions without negative consequences; and

WHEREAS government inspection does not guarantee safe, clean food and in fact is being discredited and disparaged as ineffectual in many quarters; and

WHEREAS rural electrification, refrigeration, stainless steel and communication render inappropriate decades-old comparisons and reduce the threat of food-borne contamination from small scale, rural production/processing entities; and

WHEREAS consumers who so desire should have the freedom of choice to patronize local farmers; and

WHEREAS a direct relationship link between producer and consumer best facilitates honest, moral, ethical business practices,

Be it therefore resolved that all agricultural products processed and marketed on or within 40 miles of the producing farm shall be exempt from government inspection up to and including the following annual limits: 20,000 head of poultry (current); 300 beeves; 500 hogs; 1,000 sheep, lamb and goats; 50,000 gallons of milk, 50,000 pounds of cheese, yogurt, butter and ice cream.

WHAT IS THE DIFFERENCE?

POLYFACE BEEF	CONVENTIONAL BEEF
No vaccinations	Vaccinations (immuno-depressant)
Once-in-a-blue moon medications	Routine medications
Late Spring calving--when deer are fawning	Winter calving--stress
Shed feeding to protect ground and hay--winter	Damaging winter feeding--ground tromping and impaction; soiling of feedstuff
Winter feeding of perennial polycultures	Winter feeding of hay + annuals--require tillage and use chemicals to grow
Deep carbonaceous bedding for lounging area, protected from rain (leaching) by roof	Winter nutrient pollution--leaching of manure and urine into water
Sanitary winter housing--deep bedding at proper carbon/nitrogen ratio is warm, clean and produces natural antibiotics.	Unsanitary winter housing--mud, manure, and/or concrete

WHAT IS THE DIFFERENCE?

POLYFACE BEEF	CONVENTIONAL BEEF
Winter manure protected and composted before field application--pathogens destroyed	Winter nutrients unprotected and/or applied raw--pathogens reinfest field
Sanitary winter feeding--in V-slotted feedergate excluding animal soiling of feed (hay)	Unsanitary winter feeding--on ground, in animal-soiled feeders
Compost and natural soil amendments	Chemical fertilizers
Controlled grazing--stimulates polyculture and high forage succession (prairie)	Continuous grazing--weakens good species; encourages weeds and brushy plants
Nutrient allocation for beneficial assimilation--shademobile	Nutrient translocation--shade trees, streams, campsites (cow lounge areas)
Limited or complete prevention of access to riparian areas--pump out into portable tank	Generally unlimited access to ponds and streams for water
Natural wormers--Shaklee Basic H soap and intensive rotational grazing for paddock rest	Systemic grubicides and wormers--permeate muscle and organs

WHAT IS THE DIFFERENCE?

POLYFACE BEEF	CONVENTIONAL BEEF
Sanitary pasture--free range chickens (Eggmobile) sanitize paddocks before regrazing by scratching through cow paddies (eat fly larvae and exposure of manure to sunlight)	Unsanitary pasture--regraze pathogen/parasite-infested zones (constant animal access feeds pathogens)
Forest fenced out and steep hillsides reforested--biodiversity encouraged (creates more stability in the ecosystem)	Inappropriate grazing areas--forest and steep hillsides--biodiversity discouraged
Often never see headgate in whole life--no stress	Numerous and routine corral/headgate experiences--stress
Close human contact--quiet, contented animals	Sparing human contact--"moving" becomes a major chore (stress on cow and farmer)
No hormones	Anabolic steroids for faster weight gain
Kelp meal--dehydrated seaweed	Synthetic minerals/vitamins
Pasture paddocks--moved every day	Feed lots--smell, filth

WHAT IS THE DIFFERENCE?

POLYFACE BEEF	CONVENTIONAL BEEF
Forage feeding--permanent ground cover; no chemicals or tillage	Grain feeding--expensive, erosive, ecologically unsound
Ecologically enhancive--70 percent of all grain grown in America goes through multi-stomached animals (this is unnecessary and accounts for the majority of all pesticide, herbicide, chemical fertilizer and agricultural petroleum use)	Ecologically destructive--each bushel of corn costs two in bushels of soil; irrigation depletes water resources and causes salinization of soil
Runs on solar energy	Runs on petroleum--costs 15 calories of energy to get 1 calorie of food
Forage fattened--fat outside muscle and lower saturated fat (yellowish)	Grain fattened--fat inside muscle (marbling) and higher saturated fat
Local transportation--produced, processed and sold locally	Long distance transportation--1000 miles for average steak in U.S.
Rich taste	Bland taste

WHAT IS THE DIFFERENCE?

POLYFACE BEEF	CONVENTIONAL BEEF
No chlorine carcass baths	Chlorine carcass baths
Decentralized food system	Centralized food system
Consumer/producer relationship	Consumer/producer alienation
Edible	Inedible

ORDER BLANK - SPRING 1995

POLYFACE, INC.

CHICKEN $1.35/lb.

Number of birds wanted	Processing dates
________	May 16, 17, 19, 20
________	June 13, 14, 16, 17
________	July 5, 7, 8
________	August 15, 16, 18, 19
________	Sept. 5, 6, 8, 9
________	Sept. 26, 27, 29, 30
Chicken liver	NO EXTRA

STEWING HENS (available only Sept. 26, 27, 29, 30) $1.25/lb.

__________ How many?

READY-TO-LAY PULLETS (will begin laying in September) $5.00 apiece

__________ How many?

[continued]

SALAD BAR BEEF
Available in late October.
All prices are shipping and handling charges based on hanging weight, per pound and include cut, wrap and freeze. If we happen to have one in the Spring, indicate your desire to get it earlier

_______ Split half $1.95 85-150 lbs.

_______ Half $1.90 170-300 lbs.

_______ Whole $1.85 350-650 lbs.

_______ Do you want liver?

_______ Ground quarter $1.45 100-150 lbs.

Beef take-home weights are 25-35 percent less than carcass hanging weight due to trimming and de-boning. Ditto pork.

RABBIT (available year-round) $2.75/lb.

___________ How many?

We will automatically match up rabbit with other things. For multiple pickups, indicate when.

PORK (all fresh) Prices are just like beef.

July	October
___ Half $2.05 (80-100 lbs.)	___ Half $2.05 (125-150 lbs.)
___ Whole $2.00 (160-200 lbs.)	___ Whole $2.00 (250-300 lbs.)

Comments, questions, suggestions welcome on back. Document this order somewhere so you can remember what you just did. October is months away. *[Of course, original form had spaces for name, address and phone number.]*

Spring 1995

Dear folks:

Thank you for voting. Isn't it great that you can vote every time you put food in your mouth? In a day when the easiest thing to do is complain, regulate, join an organization or vegetate and procrastinate, YOU are proactive, going right to the greatest voting booth -- the pocketbook -- and registering your choice. To the question "What can I do?" you respond: "Vote with your pocketbook and your lifestyle. You can do it every day, not just in November."

We are impressed, humbled, and challenged by your aggressive, paradigm-changing actions. Every time you patronize Polyface, Inc., you vote for:

1. Environmentally enhancive agriculture.
2. Bioregional food sufficiency.
3. Seasonal production cycles.
4. Decentralized food systems.
5. Humane animal husbandry.
6. Entrepreneurial private sector small business.
7. Relationships between rural and urban.
8. Rural non-industrial economic development.
9. Biodiversity and soil building.
10. Family friendly agriculture.
11. Home cooking instead of processed food.
12. Clean, nutritious, personally-inspected food.
13. Nonembarrassing farm incomes.
14. Emotionally exhilarating lifestyles.

Go ahead, give yourself a pat on the back, you deserve it. You have opted out of devitalized supermarket fare, taken personal charge of your food, and made a moral, cultural, social, political statement that offers hope instead of despair, action instead of frustration, and purpose instead of dismay. Instead of demanding salvation by legislation, you exemplify the true American spirit - the freedom to make right choices and do the right things. We honor you, we congratulate you, and we thank you. Carry on.

APPRENTICESHIP PROGRAM

For several years now we have turned down many requests from all over the world for apprenticeships. People want to come and learn about profitable and ecological agriculture. We are moving in a cottage for two apprentices this year. One is from Virginia and the other from Mexico. We have a waiting list of young people who want to come for extended periods (6-12 months). A milestone for us, this will facilitate comprehensive hands-on teaching and provide additional labor. With approximately 4-6 out-of-state parties a week visiting the farm, our time is becoming more and more precious. Do any of you have a decent, functional electric range/oven for this cottage?

This winter we spoke in agricultural conferences in Missouri, Kansas, Nebraska, Colorado (twice), New Mexico, Tennessee, and Pennsylvania. In addition, Joel is doing some radio programs via telephone hookup. People are starved for our message of agricultural truth. Apprenticeships are not only consistent with our educational philosophy, but will take some of the day-to-day labor pressure off, freeing us to do more research and development, teaching and writing. In direct opposition to most apprenticeship programs, however, we wanted one that would pay well and offer a business opportunity somewhat independent of the mother enterprise. That brings us to the other half of this equation.

PASTURED EGGS

Big news, big news. In the past, our eggs have always been a byproduct of our sanitation program: the Eggmobile behind the cows; the Raken house birds under the rabbits. They never had to pay their own way. Due to a series of events, last year we ended up with extra layers and consequently extra eggs. One of our longtime loyal patrons noticed the problem and, on his own volition, took some up to a couple of gourmet restaurants in Washington, D.C. At the risk of sounding proud, let's just say they were VERY well received. Of course, he could sell a hat rack to a moose. Because of the tremendous opportunity, we are now in the egg business. We've built two 120 ft. x 20 ft. hoop greenhouses to provide winter comfort for both birds and caretakers and right now we are moving 900 layers (primarily nonhybrid Rhode Island Reds) out onto pasture in specially-designed pens similar to the broiler pens.

The apprentice will receive a handsome salary (we've never apologized for desiring a professional's salary) from running the egg program and the goal is to be at 100 dozen per day by early fall. That is the volume we need for us, the apprentice, and the marketer/deliverer to get a decent return to labor. We've watched our egg knowledge grow astronomically these last six months, and the quality shows it. If this all continues the say it is going, by next year we will have another tried and proven prototype that can offer a white collar salary from a pleasant life in the country on less than 20 acres. The foundation, of course, is a product unequaled in the whole world. We'd love to tell you how these chefs are reacting and what they are finding, but space will not permit.

Because of the eggs now having to pay their way, and the tremendous demand from chefs, effective immediately our price is $1.50 per dozen at the farm, $1.00 for pullet eggs, and 50 cents for cracks. Delivered price is $2.40. A dozen eggs weighs roughly 1½ pounds. Our chicken is $1.35 per pound, beef $1.90 per pound and pork $2.00 per pound (these are competitive retail prices). Clearly at $1.50 eggs are a protein bargain and you need not worry about cholesterol in chlorophyll-rich, fresh eggs. Remember, we use the term "You get what you pay for" for stereo equipment, automobiles and clothing; it also applies to food. Nutrition, taste, and purity are directly linked to craftsmanship. The cheapest food is generally not the best buy.

We would love to have eggs available in the Staunton area so you would not have to drive out to the farm to get them. If you know of a business location that would be willing to let us install a refrigerator/honor system, please let us know. Otherwise, come on out -- these eggs will keep for a couple of months because we are moving everything out each week. Get together with a couple neighbors and trade trips. Our eggs are as much better than what's in the store (including EB eggs) as our broilers are. Gourmet chefs wouldn't lead you wrong.

PASTURED RABBIT

Daniel is on the exponential side of the learning curve. Everyone is raving about these pastured and forage-raised rabbits. For the first time, he was able to fill all the orders last year and has doubled his stock to handle the increased demand. Remember that this offers the highest protein of all meats without saturated fat because a rabbit only stores fat in its viscera. Because of its density, a pound is more filling than a pound of anything else. If you want to try some but are uneasy about cooking instructions, we'll help you. He should have about 50 ready the middle of April, before the first batch of chickens, so if you are out of meat now and would like to try some, let us know. Otherwise, we plan to coordinate rabbit pickup with chicken, beef and pork to save you extra driving, especially people from outside the Staunton area.

PIGAERATOR PORK

The response to our pork has been, in modern youngster lingo, "Awesome." As you know, our "Pigaerator Pork" serves two functions: making aerobic compost out of the winter cattle bedding pack, and cultivating. Expanding on our "Tenderloin Taxi" R&D project last year, we will attempt a three-year conversion of two acres from forest to pasture using pigs as a tillage tool. Small grains, corn, cover crops, smother crops and green manures will be harvested and incorporated using pig power to eliminate the tractor and bushhog in this process. Biodiversity calls for pastures in forests as much as forests in pastures. To stimulate biodiversity, therefore, we are attempting to reopen some of the flat "sods" in the forest that the Indians and buffalo maintained with fire and grazing. We apologize for not having enough last year. We thought a four-fold increase would be plenty, but it wasn't. We plan to double this year. No price increase.

PASTURED POULTRY

As conventional factory poultry farming becomes more and more ridiculous, our positive alternative offers a sharper contrast. The differences truly are amazing. Since the poultry industry degrades water, air, people and food, what a relief to know that we can enjoy a positive alternative. As usual, we'll offer seven batches throughout the summer, first come first served. When the last batch fills up, we'll move you to the next closest one, working both ends toward the middle until it fills up.

We encourage you to actually circle the tentative processing day on the order form. That will save us a call; you will receive a postcard reminder. If you don't feel comfortable making tentative specific dates now, don't worry about it and we'll call you a week in advance like normal. Please circle the liver question. Some of you LOVE liver and some of you HATE liver. We want to make the right matchups, so please help us here. Price same as last year.

One other thing. We use now nearly 100 5-gallon plastic buckets in our pastured poultry endeavors. Please, please keep your eyes open for these buckets and bring them with you (we'll barter for them at $1.00 per bucket) or call us and let us know where there's a stockpile. Although we try to keep each one as long as possible, they do break and need to be replaced periodically. They do not need to be clean and it does not matter what was in them or what color they are. We have a cleansing rotation that works well. Don't forget - 5-gallon buckets.

PASTURED STEWING HENS

These are our older hens who have completed their second laying season. We should have about a hundred this year. The taste is rich and meat needs to be cooked long and slow to insure tenderness. Broth is incomparably superb. Same price as last year.

SALAD BAR BEEF

Since so many of you have asked for small portions, last fall we took two down to T&E Meats in Harrisonburg and had them processed under federal inspection so we could sell them by the cut out of our freezer. Some of you have already been by to pick up some, and we encourage the rest of you to come on out and help yourself. It will be here as long as it lasts. We have ground (2.50), roasts (3.50) and steaks (4.00). Enjoy.

Our regular fall beef is the same as always, just getting better. As pasture fertility increases, we're enjoying seeing the beef become more succulent. Rather than taking time here to repeat all the ways it is different from conventional fare, we'll be glad to detail this for anyone who asks. Remember that if the portions are too large for you, inviting a friend to "go in with you" is a great way to solve the problem. Just so you'll know, our prices are a good 10 percent BELOW the price of an equivalent volume purchased by the cut from retain meat counter. That adds up to a savings of $20-40 per side. Same price as last year.

You will notice some jargon on the order blank that sounds strangely bureaucratic -- it is, to satisfy the feds. Indicate if you'd like a smaller or larger size.

READY-TO-LAY PULLETS

Again, we encourage you to produce your own eggs in your own backyard. The joy and satisfaction of producing some of your own food is incomparable. It's a great family activity. If you use a portable pen like we do, you do not need a chicken house or chicken yard. Just a little backyard and the pen is all you need -- and you'll have the greenest yard in the neighborhood. The hens are extremely quiet: it's the roosters that make the ruckus (just like a man, huh?). These are heavy breed, nonhybrid birds.

POLYFACE FIELD DAY IV, AUGUST 5, 1995

Many of you may remember 1992 when we had 350 people from 24 states for our third field day. This year we'd like to see

1000. It's being co-sponsored by STOCKMAN GRASS FARMER (a magazine for which Joel writes a monthly column) to take the correspondence pressure off of us. We'll have pigaerator pork barbecue, pastured poultry barbecue and all the fixings. If you'd like to experience the heartbeat of eco-farming (economical and ecological), you are all invited at HALF price. We will tour the farm, discuss philosophy and practice, fellowship and enjoy some of the finest food this side of heaven. Half price is $19.50 for adult, $35.00 per couple, $9.75 for 5-12 year-olds and under 5, free. MArk this date on your calendar and let us know as soon as possible about your attendance so we can procure tickets from STOCKMAN. This is a special deal we've negotiated just for Polyface patrons. It will do you good to meet folks from all across America who refuse to bow to genetic engineering, chemical agriculture, inhumane livestock models and agricultural serfdom, but who instead are implementing creative, dynamic natural models, romancing their children and communities into healthful food opportunities. Join us.

THIS AND THAT

You will no doubt see visitors working with us throughout the summer. In an effort to meet the demand for information we are collaborating with a couple of organizations who have asked us to provide hands-on teaching to folks who want to duplicate some of our methods. Occasionally these folks may ask you why you patronize Polyface -- many of them are scared to death that people in their community will not respond to exceptionally good food. Please help them over their fears and be an evangelist for this type of farming in other communities around the country.

Want a fine dining experience? Try JOSHUA WILTON HOUSE in Harrisonburg (703-434-4464). They use our eggs, rabbit, chicken (in season), pork and beef, as well as locally grown clean vegetables and berries. Fabulous.

GARBAGE RECYCLING

Many of you may not have earthworm barrels, compost piles or backyard chicken flocks to recycle your kitchen scraps (food garbage). This is wonderful organic material that should never go in a landfill (most of it does). This year we will have a couple of barrels out near the parking area where you can dump this garbage. PLEASE: no paper, no plastic, no metal. We can't afford to pick through it. Keep it in a sealed plastic bucket and dump it when you get here -- you can rinse out the bucket here. Or you can bring it in a plastic bag, but you must take the plastic bag home. This is a trial, but we think you are all "tuned in" enough to realize a great recycling opportunity. We plan to feed these scraps to the chickens and convert them into eggs (and save some purchased feed costs). You may get your neighbors involved as well. We think we can handle quite a bit. Let's see if we can creatively turn more liabilities into assets. Thanks.

As always, it's been a real pleasure visiting with you. To keep this from being a one-sided conversation, share your comments, criticisms and encouragements on the back of the order form and in person when we see you during the year. Be sure to post this letter and/or whatever decisions you make so you'll remember what you ordered. Thank you for your interest and support, and we can't wait to see you. Blessings on you.

The Salatins of Polyface, Inc.

Appendix B

Resources

MAGAZINES

ACRES USA
P.O. Box 8800
Metairie, LA 70011
(504) 889-2100

Countryside and Small Stock Journal
N2601 Winter Sports Rd.
Withee, WI 54498
(800) 551-5691

Permaculture Activist
P.O. Box 1209
Black Mountain, NC 28711
(704) 683-4946

Small Farmer's Journal
P.O. Box 1627
Sisters, OR 97759
(503) 549-2064

Stockman Grass Farmer
5135 Galaxie Drive
Suite 300C
Jackson, MS 39206
(800) 748-9808

American Small Farm
9420 Topanga Canyon
Chatsworth, CA 91311-5759
(818) 727-2236

Holistic Resource Management Quarterly
1007 Luna Circle NW
Albuquerque, NM 87102
(505) 842-5252

Quit You Like Men
P.O. Box 1050
Ripley, MS 38663-9430
(601) 837-4596

Small Farm Today
3903 W. Ridge Trail Rd.
Clark, MO 65243-9525
(800) 633-2535

BOOKS

Berry, Wendell
The Gift of Good Land
The Unsettling of America

Bromfield, Louis
Out of the Earth
Pleasant Valley

Davis, Adelle
Let's Cook it Right

Doane, D. Howard
Vertical Diversification

Faulkner, Edward H.
A Second Look
Plowman's Folly
Soil Development

Howard, Sir Albert
An Agricultural Testament

Jackson, Wes
Altars of Unhewn Stone

Leatherbarrow, Margaret
Gold in the Grass

Mollison, Bill
Permaculture One
Permaculture Two

Murphy, Bill
Greener Pastures on Your Side of the Fence

Nation, Allan
Grass Farmers
Pasture Profits with Stocker Cattle

Rodale Staff
The Complete Book of Composting
The Rodale Book of Composting

Savory, Allan
Holistic Resource Management

Smith, Burt
Intensive Grazing Management: Forage, Animals, Men, Profits

Staten, Hi W.
Grasses and Grassland Farming

Turner, Newman
Fertility Pastures and Cover Crops

Walters, Charles, Jr.
An ACRES USA Primer
Weeds, Control Without Poisons

Thompson, W. R.
The Pasture Book

Voisin, André
Grass Productivity

Index

A

B

C

D

E

F

G

H

J

K

L

M

N

O

P

Q

R

S

T

U

V

W

Y